Multi-Sensor and Multi-Temporal Remote Sensing

Specific Single Class Mapping

Anil Kumar,
Priyadarshi Upadhyay and
Uttara Singh

CRC Press is an imprint of the
Taylor & Francis Group, an **informa** business

First edition published 2023
by CRC Press
6000 Broken Sound Parkway NW, Suite 300, Boca Raton, FL 33487–2742

and by CRC Press
4 Park Square, Milton Park, Abingdon, Oxon, OX14 4RN

CRC Press is an imprint of Taylor & Francis Group, LLC

ISBN: 978-1-032-42832-1 (hbk)
ISBN: 978-1-032-44652-3 (pbk)
ISBN: 978-1-003-37321-6 (ebk)

DOI: 10.1201/9781003373216

Typeset in Times
by Apex CoVantage, LLC

Multi-Sensor and Multi-Temporal Remote Sensing

Specific Single Class Mapping

This book elaborates fuzzy machine and deep learning models for single class mapping from multi-sensor, multi-temporal remote sensing images while handling mixed pixels and noise. It also covers the ways of pre-processing and spectral dimensionality reduction of temporal data. Further, it discusses the 'individual sample as mean' training approach to handle heterogeneity within a class. The appendix section of the book includes case studies such as mapping crop type, forest species, and stubble burnt paddy fields.

Key features:

- Focuses on use of multi-sensor, multi-temporal data while handling spectral overlap between classes
- Discusses range of fuzzy/deep learning models capable to extract specific single class and separates noise
- Describes pre-processing while using spectral, textural, CBSI indices, and back scatter coefficient/Radar Vegetation Index (RVI)
- Discusses the role of training data to handle the heterogeneity within a class
- Supports multi-sensor and multi-temporal data processing through in-house SMIC software
- Includes case studies and practical applications for single class mapping

This book is intended for graduate/postgraduate students, research scholars, and professionals working in environmental, geography, computer sciences, remote sensing, geoinformatics, forestry, agriculture, post-disaster, urban transition studies, and other related areas.

When on a joke we laugh once. . . .
Than why we cry number of times on a problem. . . .

Life is Fuzzy. . . .
Use Fuzzy Logic. . . .
. . . . Remain Crisp. . . .

Remember. . . .
Life Once. . . .
Birth Once. . . .
Death Once. . . .

To
Gurus who guided us,
Students who worked with us,
and
Readers of this book

Have Right Choice. . . .
Nothing Is Impossible. . . .
I Am Possible. . . .
Impossible. . . .

Contents

. . . . On top of every engineering. . . .
do human engineering with specialisation in spirituality (aadhyaatmikata)

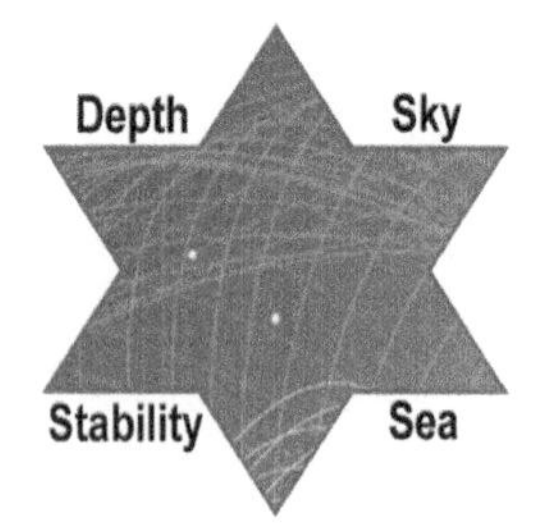

Foreword

The scientific community is actively engaged in developing solutions to the multiple, grave issues that confront us, including food and water security, resource management, environmental protection, sustainability, disaster resilience, and climate change. One modern tool that has come to occupy centre stage is the use of Earth observation (EO) data. In the short span of five decades since EO data became available, the various technologies that support it have seen multiple evolutionary transitions. Currently, EO data is available daily from hundreds of satellites covering the entire globe multiple times in a day in various electromagnetic regions and spatial resolutions. The availability of open EO data, a turn-around-time of a few hours, pre-processed, ready-to-use data hosted on the web, and the geospatial web access of current data along with historical time series have created a gap in tools for accurate, quick, and ready-for-use information from EO data.

Traditional EO data analysis has stressed pattern recognition, change analysis, and ingesting EO information in GIS, in decision-support systems, or as quantitative parameters in models. There has been an increased use of multi-sensor, multitemporal data with tools to answer specific 'single-class' information needs, in contrast to a generic and very detailed discrimination and mapping of an entire study area. Some of the examples include discrimination of a specific crop, discrimination of a specific species in the forest with hyperspectral data, and identifying invasive species in forests/natural communities (current).

In this context, this co-authored book by Anil Kumar, Priyadarshi Upadhyay, and Uttara Singh, titled *Multi-Sensor and Multi-Temporal Remote Sensing: Specific Single Class Mapping*, is very timely and useful. After a short introduction to EO data and spatial spectral-indices, the book introduces the concepts of single class mapping, including the use of a fuzzy approach with an emphasis on the approach to the training of classifiers. The last two chapters introduce various machine-learning algorithms. The utility of the book is enhanced by two appendices that present seven real-world case studies and also describe a processing tool, SMIC, that has been developed and improved by the senior author over last nearly two decades.

I compliment the authors for providing material for use by researchers and practitioners of most modern tools which will help the user community to better utilise vast and ever-expanding EO data, as well as to spur the EO researchers in their research for establishing new and more efficient discrimination techniques.

Dr. Vinay Kumar Dadhwal
Indira Gandhi Chair Professor of Environmental Sciences
National Institute of Advanced Studies (NIAS)
Bengaluru—560012, INDIA

Classification of remote-sensing images is a longstanding activity. It has been a necessity since the collection of the first images decades ago. Scientists and users of images have been wondering about what the images represent and what on the

Earth's surface can be seen from space. During the many years that satellites have been operating and images have been collected, the satellites and their images have experienced several major changes. New satellites were launched that were geostationary, polar, or other, and that had alternate spatial resolutions. The traditional 30 x 30 m resolution, as of the Landsat images, is still relevant, but both finer and coarser resolutions have found their places. Also, the frequency of image capturing has changed, with satellites currently having a return time of a few days instead of several weeks. Further, a wide range of optical and radar sensors are now operating onboard the satellites. Optical sensors are commonly multi-spectral or hyperspectral, while radar images may result into InSAR images, polarimetric or not. Most importantly, however, the questions about what to observe from the Earth's surface have changed. Initially, individual band information was sufficient, and after pre- and post-processing, any required information could be derived from the images. More recently, advanced products have been delivered by space organisations. In this way, the bridge between the images as the product and the information for the users was shortening. In all these developments, great scientific challenges were made to make the most out of the collected images. Such methodological aspects with a strong statistical and computer-science background have survived the changes in technologies while being adjusted to the latest modifications.

The current book provides the necessary introduction and overview of the latest developments in this field of image classification. Dr. Kumar as a well-known expert in the domain of image classification has been a prominent leader in developing the domain of classification and the related fuzzy classification methods with intensive research activities. Related to his network of eminent co-authors, these activities have resulted in the volume *Multi-Sensor and Multi-Temporal Remote Sensing: Specific Single Class Mapping* that is available at this moment. In this book, it is prominently realised that much of the world is not crisp, but rather, is of a gradual and fuzzy nature. In fact, if one thinks about it, sharp boundaries are generally non-existent on the globe. Fuzzy theory provides an excellent set of methods to address this aspect. Fuzziness is general, and the focus of the current book—'Specific Single Class Mapping'—is the single most important topic in the end; to have a final classification of each entity, where the detailing of the class—i.e. the legend—can be complicated.

At this place, I am happy to complement Dr. Kumar with this achievement. I most heartily recommend the book to all who are interested in identifying fuzzy information from space by classifying fuzzy objects from remote-sensing images into single classes, including the associated uncertainties.

Prof. Dr. Ir. Alfred Stein
Faculty of Geo-Information Science and Earth Observation (ITC)
University of Twente
The Netherlands

I am, indeed, very happy to write this foreword to the book *Multi-Sensor and Multi-Temporal Remote Sensing: Specific Single Class Mapping*. Ever since satellite-based remote sensing has found a place in data collection and analysis, it has created a huge impact on human life. Natural resource and disaster monitoring has benefited many folds through the usage of satellite data. In the earlier 1980s, remote sensing was restricted to 4 spectral band—i.e. 3 visible and 1 NIR band. Slowly, the user committee witnessed an increase in the number of spectral bands in which data was being collected. Side by side, the spectral resolution and spatial resolution started to improve. The quest to get more information led to the finer division of spectral band-width. This brought about the introduction of hyperspectral remote-sensing. The band width was much finer then that which was available in the earlier sensors.

In the initial days, the United States of America had the sole capability of collecting Earth-related data. In the 1980s, France and India joined the league. Thereafter, many more countries started to acquire satellite-based data. New sensors were introduced with enhanced capability in spectral, spatial, and radiometric resolutions. This brought about a new dimension in analysis—i.e. multi-sensor—into the analysis domain. Further, more missions came into existence, bringing about another dimensional change, leading to the reduction in re-visit frequency or temporal resolution.

With these improvements in the four basic resolutions in satellite data, the analyst started looking at multi-sensor and multi-temporal-based analysis. The advantages of this concept led to improvement in the monitoring of events in agricultural domains such as crop monitoring, pest infestation, and improved forecasting of crop yield. Further, the need to assess damage due to various natural phenomena such as drought, floods, landslide, and snow avalanche received a tremendous boost. The climate-change phenomenon is expected to be understood better by using different types of satellite data having different spectral, spatial, radiometric, and temporal resolutions.

I am indeed happy that the authors have thought of penning down the knowledge and expertise by writing this book. It will allow the young researchers and seasoned analysts to understand the fundamentals of such a type of analysis. One important aspect of the book is the focus on quick identification and extraction of a single class of information using knowledge-based techniques of machine learning. The authors have a good background in this area and the book is a true reflection of the same.

As guru to the authors, I can only bless them and wish that they should continue to achieve greater heights and continue to disseminate their knowledge through their books.

Prof. Sanjay Kumar Ghosh
Former HOD and Professor of Civil Engineering
Indian Institute of Technology Roorkee
Roorkee—247667, India

Earth observation systems are generating petabytes of data every day and extracting the knowledge from the Data Mountains is a huge challenge. On top of it, some of the applications require that the data are collected over a period of time at regular or irregular intervals and analysed together to understand the spatio-temporal patterns reflected in the multi-date and multi-sensor datasets. This is the context in which the upcoming book *Multi-Sensor and Multi-Temporal Remote Sensing: Specific*

Single Class Mapping by Dr. Anil Kumar, Dr. Priyadarshi Upadhyay, and Dr. Uttara Singh attempts to expose students, researchers, and professionals to the science and tools related to multi-temporal Earth observation.

The material is covered in six concise chapters, starting with an introduction to multi-temporal, remote sensing in the first chapter, followed by various indices in Chapter 2. Chapter 3 introduces multi-sensor, multi-temporal remote sensing in which various temporal indices are presented. Chapter 4 discusses an interesting and important topic related to training data collection, strategies for collecting data for heterogenous and homogeneous classes, application-specific data collection, etc. Chapter 5 covers fuzzy and possibilistic classification algorithms for a specific one-class classification and mapping (one class of interest versus everything else as background). Fuzzy set-theoretic methods are quite powerful but there are not enough books, particularly in the remote-sensing arena, covering the concepts, algorithms, and geo-spatial applications. This book fills an important gap that exists related to this area.

Chapter 6 presents many of the popular machine-learning tools for temporal data inputs, such as Recurrent Neural Networks (RNN), Long Short-Term Memory (LSTM), Gated Recurrent Unit (GRU) and Convolutional Neural Network (CNN). Two appendices help the reader to understand the application by including several case studies and a software tool, SMIC, developed at the authors' institution. It can safely be said that there are hardly any books in the remote-sensing application area discussing these modern, multi-temporal data-analysis methodologies and this book is a welcome addition for the reader community.

Many applications that have an influence on our lives—such as disaster management, crop yield, forest vegetation-health monitoring, urban growth, and coastal erosion—require repeated observation of our Earth at intervals driven by the needs of the applications. Luckily for the end user, there are now several Earth Observation systems that are providing data completely free (e.g. Sentinel series of multispectral and synthetic aperture radar images, Landsat multispectral images), low-cost yet very useful data sets (e.g. Resourcesat and Cartosat), and very high-resolution and expensive data sets (e.g. Worldview—2/3, Pleaides, Planet Labs, Maxar, Radarsat) at regular intervals or on order basis. The challenge today is not so much in availability of data or processing power, which was the challenge 20 years ago, but in strategies to efficiently analyse the multi-sensor/multi-temporal image datasets and provide the administrators and end-users with usable inputs for informed decision-making. This book addresses many of these issues in this new book by the authors who are experts in the field of remote sensing, machine learning, and use of geo-spatial technologies for critical applications such as vegetation mapping, identification of burned and degraded vegetation, urban-growth and urban-sprawl monitoring, and so on.

Overall, this book is a very welcome addition to the multi-sensor and multi-temporal remote-sensing processing literature, and I recommend that it is seen on the bookshelf of every practitioner/student. Given the expertise of the authors in remote sensing, it is expected that the practitioners of multi-sensor and multi-temporal remote sensing related image/data analysis will benefit highly from using this book, but the tools and techniques presented in the book are equally applicable in other domains and are therefore beneficial to students and researchers from those domains.

Dr. Krishna Mohan Buddhiraju
Professor of Centre of Studies in Resources Engineering,
IIT Bombay, India

'ask not what your country can do for you, ask what you can do for your country'
....John F. Kenndy

A River Disappear to Become Ocean....

Detach for All in You and
You Belongs to Everyone....

Preface

A massive amount of remotely sensed data has been collected in recent years as a result of the availability of various Earth observation satellite sensors. The principal application of remote-sensing data is the creation of thematic maps.

However, one of the most important uses is a higher-level, specialised-class mapping, together with temporal information. While mapping a single or specific land cover/class, the spectral overlap is a practical issue to be taken care of. This issue can be handled by employing the temporal, remote-sensing images or data. Cloud cover is another problem while opting for optical, remote-sensing-based data to make a temporal database. This problem can be resolved by using a dual- or multi-sensor technique and by having a sufficient amount of temporal data. Further, in the present scenario, there are some machine-learning models that are more advanced and capable for mapping a single, specific class of interest. Special crop-mapping, disaster-affected region mapping, specific forest-species mapping, burnt paddy field mapping, and so on, are examples of specific-class mapping.

While extracting class information from remote-sensing data using conventional, hard classifiers, some of the issues—such as the mixed-pixel problem, non-linearity between classes, and noisy pixels as unclassified pixels—remain to be handled.

Application of fuzzy machine-learning algorithms can provide highly realistic classification results while mapping a specific class of interest. Fuzzy machine-learning algorithms can deal with mixed pixel problems using a soft computing approach while the non-linearity across classes can be handled using the kernel concept. Thus, a single class of interest can be identified using a learning algorithm and probabilistic-based fuzzy classifiers. Though the fuzzy classifiers are independently capable of solving the mixed-pixel problem, they cannot solve the problem of heterogeneity inside a class. Similarly, in a non-homogenous training sample, the statistical-based classification models using statistical parameters cannot handle heterogeneity.

Thus, with an increased use of satellite-based Earth-observation data, there is a great deal or potential to combine diverse remote-sensing images to extract a lot of unique information. There are many applications that demand database information regarding to single, unique class. For example, a single, specific class in the form of unique crop-acreage provides useful information for determining the area of a certain crop. Thus, the crop-production estimation and crop insurance claim application can both benefit from such specific crop-area information. Mapping of individual species in a forest can be used in forestry and ecology applications. Further, the remote-sensing images are also useful for estimating and mapping the post-disaster damage.

Fortunately, the explanations of these single, unique class-mapping methods have shown to be suitable for graduate, postgraduate, and research students and academicians from a varied user population. Furthermore, the methods described in the literature do not provide straightforward solutions for integrating the multi-sensor, remote-sensing images to map a single, specific class of interest having spectral overlap with other classes in a particular area.

The goal of this book is to consolidate information in the form of fuzzy machine- and deep-learning models for single-class mapping from multi-sensor, multi-temporal remote-sensing images in one place. The necessity of multi-sensor, temporal remote-sensing data for specific land cover/class mapping is explained in this book. The chapters of this book are arranged to give reader an introductory understanding about the capabilities of multi-spectral, hyper-spectral and SAR remote-sensing data with more emphasis on the use as well as processing of multi-sensor and multi-temporal remote-sensing data.

Chapter 1 of this book provides information about the capabilities of multi-spectral, hyperspectral and SAR images, as well as importance of dimensionality reduction. Chapter 2 deals with various spectral and texture-based indices with an emphasis on using them while processing the temporal data and hence generating the temporal-indices database. While generating the temporal-indices database, the role of indices is to reduce the spectral dimension and keep temporal information.

Further, to reduce the spectral dimension, in addition to the conventional approach, another new Class-Based Sensor Independent (CBSI) has also been included in Chapter 2. The purpose of the book is to cover the state-of-the-art fuzzy machine- and deep-learning models to map a single, specific class of interest from single-, dual-, or multi-sensor, remote-sensing images. Before applying fuzzy machine- and deep-learning models, the single-, dual-, or multi-sensor, temporal sensor concepts have also been discussed in Chapter 3. Chapter 4 covers the role of the training-data concept and aims to understand how the fuzzy machine-learning models can address the variability within classes by using training parameters as an 'individual sample as mean' method. This chapter also talks about how to expand the size of the training sample in various circumstances from limited training data. Chapter 5 of this book deals with the fuzzy machine-learning models capable for single-class mapping, while Chapter 6 includes the capabilities of deep-learning-based models for single-class mapping. In Appendix A1 of this book, several case studies on specialised-crop mapping, burned paddy field mapping, forest-species mapping, and other topics are presented, while Appendix A2 includes an in-house, dedicated SMIC temporal-data-processing module.

Appendix A1 includes seven case studies. i) Fuzzy- versus deep-learning classifiers for transplanted paddy field mapping is one of the case studies. Other case studies include: ii) dual-sensor temporal data for forest-vegetation-species mapping and crop mapping; iii) handling heterogeneity with training samples using the individual sample as a mean approach for the medicinal crop Isabgol (Psyllium husk); iv) investigating the effect of red-edge bands on sunflower-crop mapping; v) mapping burnt paddy fields using Sentinel-2 data from two dates; vi) mapping a ten-year-old Dalbergiasissoo forest species, as well as vii) transition-building footprints.

Appendix A2 is a demonstration of SMIC's multi-temporal data-processing module: sub-pixel multi-spectral image classifier; in-house package. SMIC package has dedicated fuzzy- and deep-learning-based classifiers mentioned in this book for specific single-class mapping capability. The steps implemented in the SMIC temporal-processing package are designed to require as little human intervention as possible in order to extract specific, single classes of interest, such as specific crops, their initial sowing-stage mapping, or harvesting-stage mapping, specific species in forest mapping, stubble-biomass mapping, and application in specific post-damage-area mapping. There are provisions in the temporal SMIC tool for seed-based training-sample collections, seed-based extension of training samples, and extraction of class-spectral information to find suitable bands to be used in the CBSI (class-based, sensor-independent) approach, which is discussed in Chapter 3.

BCD in Geometry of life,
Geometry of life has two fixed points—
Birth (B) and Death (D)—in between, choice (C),
. . . that's only in our hand. . . .

Unless what is within you comes out. . . .
You cannot make an impact on yourself nor others. . . .
Sharpening which is painful brings out what is within you to make an impact. . . .

To become a masterpiece, life sculptor will give hard knock, trouble, failure. . . .
Question is whether ready to bear pain from life
sculpting tools. . . .

We are prone to make mistakes. . . .
But have an opportunity or chance to correct them. . . .
and
Learn from mistakes. . . .

Our Gratitude with three Rs

Recognising—Thankful to all those who have made a difference in our life!
Remembering all those who have done good for us—We mean it!
Reciprocating back for all the good that was done for us.
Thankful to all through three Rs of gratitude.

First of all, praise to God—the creator, destroyer, and the cherisher of the universe.

The thought in writing this book was to share practical experience about the application of multi-sensor and multi-temporal remote-sensing image-processing using fuzzy machine- and deep-learning-based classification algorithms for specific-class mapping. In this book, detailed, state-of-the-art mathematical concepts about processing multi-sensor and multi-temporal remote-sensing data for specific class mapping have been covered.

This book in your hands is the essence of the blessings of gurus, elders' support, encouragement, help, advice, and the cooperation of many people—students, friends, and family members. In the sequel that follows, some names may or may not be there, but their contributions have all been important.

The authors, having been working in ISRO centres, got inspiration, encouragement, and support from top management, starting from the chairman, ISRO, other dignitaries of ISRO (HQ), and for this we are heartily thankful.

We are also thankful to Dr. P. S. Roy, Dr. V. K. Dadhwal, Dr. Y. V. N. Krishna Murthy, Dr. A. Senthil Kumar, Dr. Prakash Chauhan, and Dr. R. P. Singh—leaders of IIRS.

Dr. Priyadarshi Upadhyay is thankful to the Director Uttarakhand, Space Application Centre, and other colleague scientists for their constant encouragement in writing this book.

We are thankful to Dr. V. K. Dadhwal for motivation to explore how to incorporate profile information of crops; he also explained how to integrate SAR data with optical data in the temporal domain, which has given thought to exploring multi-temporal as well as multi-sensor remote-sensing images. Overall, it is to be mentioned that seed for this book was sown by Dr. V. K. Dadhwal.

We are thankful to Dr. Y. V. N. Krishna Murthy for teaching about clarity and to be crisp in life.

Dr. Anil Kumar and Dr. Priyadarshi Upadhyay are highly grateful to Prof. S. K. Ghosh, Head of Department, Civil Engineering, Indian Institute of Technology Roorkee, for his constant support and ideas, as well as for the improvement of their writing skills as a supervisor during the course of their PhD research. Dr. S. K. Ghosh gave them the confidence to start their PhD research through statements like 'smallest step done differently can be research'.

The acknowledgment would not be complete without mentioning our family members who provided their love and encouragement. Dr. Anil Kumar thanks his family members, including Mrs. Geeta Kamboj, Ms. Jaya Kamboj, Mr. Arya Kamboj, Mrs.

Savita Kamboj (mother), and Dr. J. P. Singh (father), who have suffered from our absence at home during preparation of this textbook.

Dr. Priyadarshi Upadhyay thanks his family members, including his kids, Shreyaan Upadhyay and Ananya Upadhyay, for taking their precious time; his lovely wife, Nirja Upadhyay; his loving mother, Mrs. Janki Upadhyay; his father, Mr. Mahesh Chandra Upadhyay; and other family members for their every support.

For Dr. Uttara Singh, no acknowledgement would be complete without thanking the family members who provided love and care during work on this book. A special thanks to her mother, Mrs. Vibha Singh; son, Vayun Rai; siblings, Krishnarjun and Katyayini.

We sincerely acknowledge critical comments received from reviewers, which have brought the content of this book to maturity. We also thank Dr. Vinay Kumar Dadhwal, Dr. IR A. Stein (Alfred), Dr. Sanjay Kumar Ghosh, and Dr. Krishna Mohan Buddhiraju, experts in the machine-learning area, for giving their valuable introductory remarks in the Foreword.

We are thankful to Shri. Gaur Gopal Das, Indian lifestyle coach and motivational speaker, who is part of the International Society for Krishna Consciousness; he motivated us indirectly to inculcate spirituality and nourishment in this book.

Lastly, sincere thanks are due to CRC Press, Taylor & Francis Group, for publishing the manuscript as a textbook.

Anil Kumar
Head PRSD &Scientist/Engineer 'SG'
Indian Institute of Remote Sensing (ISRO)
Dehradun, Uttarakhand, India

Priyadarshi Upadhyay
Senior Scientist/Engineer
Uttarakhand Space Application Centre
Dehradun, Uttarakhand, India

Uttara Singh
Assistant Professor
CMP Degree College
University of Allahabad, Allahabad, India

Have happiness during the Journey, not at the destination. . . .

Author Biographies

Anil Kumar is a scientist/engineer 'SG' and the head of the photogrammetry and remote sensing department of Indian Institute of Remote Sensing (IIRS), ISRO, Dehradun, India. He received his B.Tech. degree in civil engineering from IET, affiliated with the University of Lucknow, India, and his M.E. degree, as well as his Ph.D. in soft computing, from the Indian Institute of Technology, Roorkee, India. So far, he has guided eight Ph.D. thesis, and eight more are in progress. He has also guided several dissertations of M.Tech., M.Sc., B.Tech., and postgraduate diploma courses. He always loves to work with Ph.D. scholars and masters and graduate students for their research work, and to motivate them to adopt research-oriented professional careers. He received the Pisharoth Rama Pisharoty award for contributing state-of-the-art fuzzy-based algorithms for Earth-observation data. His current research interests are in the areas of soft-computing-based machine learning, deep learning for single-date and temporal, multi-sensor remote-sensing data for specific-class identification, and mapping through the in-house development of the SMIC tool. He also works in the area of digital photogrammetry, GPS/GNSS, and LiDAR. He is the author of the book *Fuzzy Machine Learning Algorithms for Remote Sensing Image Classification* with CRC Press.

Priyadarshi Upadhyay is working as a scientist/engineer in Uttarakhand Space Application Centre (USAC), Department of Information & Science Technology, Government of Uttarakhand, Dehradun, India. He received his B.Sc. and M.Sc. degrees in physics from Kumaun University, Nainital, India. He completed his M.Tech. degree in remote sensing from Birla Institute of Technology Mesra, Ranchi, India. He completed his Ph.D. in geomatics engineering under civil engineering from IIT Roorkee, India. He has guided several graduate and post-graduate dissertations in the application area of image processing. He has various research papers in SCI-listed, peer-reviewed journals. He has written the book *Fuzzy Machine Learning Algorithms for Remote Sensing Image Classification* with CRC Press. His research areas are related to the application of time-series remote-sensing, soft computing, and machine-learning algorithms for specific land-cover extraction. He is a life member of the Indian Society of Remote Sensing and is an associate member of The Institution of Engineers, India.

Uttara Singh, an alumna from the University of Allahabad, Prayagraj, is presently working as an assistant professor at CMP Degree College, University of Allahabad, based in Prayagraj, Uttar Pradesh. Though being a native of U.P., she has travelled far and wide. She has contributed to numerous national and international publications, but her interests lie mainly in urban planning issues and synthesis. She is a life member of several academic societies of repute to name a few Indian National Cartographic Association (INCA), Indian Institute of Geomorphologist (IGI), National Association of Geographers (NAGI). She has also guided many PG and UG project dissertations and has guided post-doctoral research. Presently, she also holds the office of the course coordinator for ISRO's sponsored EDUSAT Outreach program for learning geospatial techniques and the course coordinator for soft-skill development programs in the same field in Prayagraj.

What is Your Life? A Candle or an Ice Cream. . . .
A candle gives light before it melts. . . .
Ice Cream is eaten before it melts. . . .

Dear Esteemed Reader:

Thank you for purchasing this book. We hope this book will enrich your understanding the concepts of multi-sensor and multi-temporal remote-sensing with processing and applications in specific single class mapping. We would be grateful to have your feedback about this book at fuzzymachinelearning@gmail.com

List of Abbreviations

AFRI	Aerosol Free Vegetation *Index*
AI	Artificial Intelligence
AIS	Airborne Imaging Spectrometer
APDA	Atmospheric Pre-corrected Differential Absorption
APEX	Airborne Prism EXperiment
ARVI	Atmospherically Resistant Vegetation Index
ASAR	Advanced Synthetic Aperture Radar
AVIRIS	Airborne Visible/Infrared Imaging Spectrometer
CARI	Chlorophyll Absorption in Reflectance Index
CART	Classification and Regression Tree
CASI	Compact Airborne Spectrographic Imager
CBSI-NDVI	Class-Based Sensor Independent-Normalized Difference Vegetation Index
CCD	Canopy Chlorophyll Density
CHI	Crop Hail Insurance
CI	Chlorophyll Index
CNN	Convolutional Neural Networks
1D-CNN	One-Dimensional CNN
2D-CNN	Two-Dimensional CNN
DDV	Dense Dark Vegetation
DIAS	Digital Airborne Imaging Spectrometer
DL	Deep Learning
dNBR	Delta Normalised Burnt Ratio
DN	Digital Number
EMR	Electro-Magnetic Radiation
ERTS	Earth's Resources Satellite
ESA	European Space Agency
FCC	False Colour Composite
FCM	Fuzzy *c*-Means
FLD	Fisher Linear Discriminant
GEMI	Global Environmental Monitoring Index
GLCM	Gray Level Co-occurrence Matrix
GRD	Ground Range Detected
GRUs	Gated Recurrent Units
IR	Infrared
IWS	Interferometric Wide Swath
$\mathbf{K_{cb}}$	Basal Crop Coefficient
KMPCM	Kernel based Modified Possibilistic *c*-Means
KMPCM-S	Kernel based Modified Possibilistic *c*-Means with constraints
LAI	Leaf Area Index
LLE	Locally Linear Embedding

LMM Linear Mixture Model
LSTM Long Short-Term Memory
LULC Land use/Land cover
MCARI Modified Chlorophyll Absorption Ratio *Index*
MCARI2 Modified Chlorophyll Absorption Ratio *Index 2*
ML Machine Learning
MLC Maximum-Likelihood Classifier
MMD Mean Membership Difference
MNF Minimum Noise Fraction
MODIS Moderate-Resolution Imaging Spectroradiometer
MPCI Multiple-Peril Crop Insurance
MPCM Modified Possibilistic *c*-Means
MSAVI Modified *Soil Adjusted Vegetation Index*
MSI Multispectral Imager
MSS Multispectral Scanner System
MTVI2 Modified Triangular Vegetation *Index 2*
NBR Normalised Burnt Ratio
NC Noise Clustering
NDBI Normalised Difference Built-Up Index
NDSI Normalised Difference Snow Index
NDVI Normalised Difference Vegetation Index
NDVIRE Normalised Difference Vegetation Index Red-Edge Indices
NDVINRE1 Normalised Difference Vegetation Index Red-Edge Indices 1
NDVINRE2 Normalised Difference Vegetation Index Red-Edge Indices 2
NDVINRE3 Normalised Difference Vegetation Index Red-Edge Indices 3
NDWI Normalised Difference Water Index
NIR Near Infrared
NPCI Normalised Pigment Chlorophyll ratio Index
OSAVI Optimised-Soil Adjusted Vegetation Index
PCA Principal Component Analysis
PCM Possibilistic c-Means
PhIX Phenological Index
PLSDA Partial Least-Square Discriminant Analysis
PVI Perpendicular Vegetation *Index*
REP Red-Edge Position
SAR Synthetic Aperture Radar
SAVI *Soil Adjusted Vegetation Index*
SPOT Satellite Pour l'Observation de la Terre
SR Simple Ratio
SVM Support Vector Machines
SWIR Short-Wave Infrared
TCARI Transformed Chlorophyll-Absorption Reflectance *Index*
TD Transformed Divergence
TM Thematic Mapper
TSAVI Transformed-Soil Adjusted Vegetation Index (TSAVI)

TVI	Transformed Vegetation Index
RMSE	Root Mean Square Error
RNNs	Recurrent Neural Network
RVI	Relative Vegetation *Index*
VNIR	Visible to Near-Infrared

When we are beautiful, it's god gift to us. . . .

When we live our life beautiful, it is our gift to god. . . .

Happiness Theorem. . . .
When divided it multiples. . . .

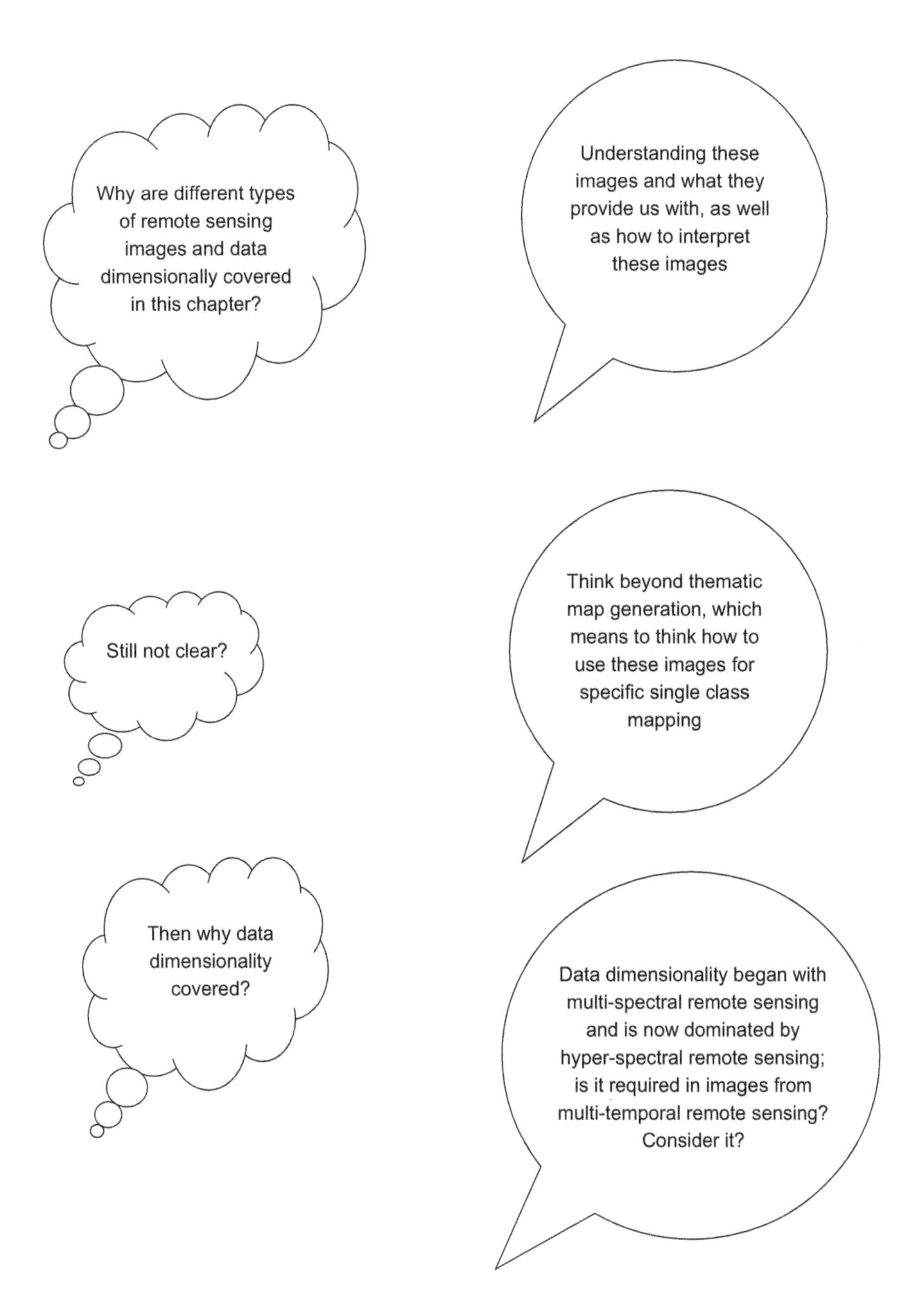

When you run alone its called Race. . . .
When God Runs with you its called Grace, To be Kind. . . .
. . . .Otherwise Success in head is Ego, Failure in heart is Depression. . . .

1 Remote-Sensing Images

1.1 INTRODUCTION

Vegetation covers almost seventy percent of the Earth's surface. Remote-sensing data can be used to extract vegetation biophysical information using different type of indices, created by various digital-image-processing algorithms (Jensen, 1996). The spectral response in the red and near-infrared regions is extensively used in plant remote-sensing. The leaf cell structure and its interaction with incident electromagnetic radiation determine this spectrum response. The structure of leaf cells varies greatly depending on the species and climatic conditions at the time of growing. In the spectral region, red-edge spectral portion for vegetation spectral signal is slope, which varies from 1% reflectance at red band to 40% slope at infrared spectral area. The slight variations in the chemical composition and morphological status of plants cause a shift of the red edge in reflectance spectra of vegetation documents (Boochs, Kupfer, Dockter, and Kuhbaüch, 1990). Furthermore, for all classification methods, the spectral behaviour is responsible for identifying the ability of extraction of crop information from a remote sensing image. Phenological variations in the crop can be better parameters for determining ground coverage. Furthermore, the presence of background soil allows for the differentiation of diverse vegetation. The amount of background soil present is also controlled by the stage of the plant's phenological cycle. Jensen (1996) established vegetation indicators for quantitative and qualitative assessment of growth dynamics and vegetation cover. It is significant at every imagined geographic scale in the biosphere.

The composition of vegetation determines the ecosystem's function and structure. Different worldwide activities, like climate and land use/land cover changes, necessitate a high-resolution, classified crop-based data (Zhang and Leung, 2004). The nutrients and fibre crops of the world are required to be gathered in a precise and up-to-date manner (Groten, 1993).Various attributes generated using spatial and spectral information of various crops are the important land-cover information, which can be extracted using satellite images (Cihlar, 2000). Plants' sensitivity to varied electromagnetic radiation is influenced by two forms of information: morphological structure and chemical composition. The primary parameters that influence spectral behaviour are different stages of plant growth, plant health, and growing environment. Between 680nm to 760nm spectral wavelength spectrum of EMR having slope is called 'red edge' (Boochs et al., 1990). The spectral curve of dry soil, green vegetation, and dead plants have extensive variations in the visible to infrared electromagnetic spectrum regions. Due to spectral variation between classes, there are evolutions of different vegetation indices (Jensen, 2010).

Crop-mapping systems which rely on visual interpretation are time consuming and qualitative in nature (Roy et al., 2015; Xie, Sha, and Yu, 2008). For thematic

DOI: 10.1201/9781003373216-1

mapping of crops, the recent technological developments have replaced old methods of digital image classification. These new developments were created to deal with spectral overlap between two or more similar classes, while mapping at level 2 or higher and hence necessitates the use of specific-class mapping. The only way to handle spectral overlap between two or more classes is to incorporate unique stages of a class using multi-temporal remote-sensing data. When temporal data is unavailable in optical remote-sensing due to cloud-cover or other issues, it is possible to use dual- or multi-sensor remote-sensing data. When there are mixed pixels in an image, soft classification techniques yield more accurate results than hard classification techniques. A soft classification method divides a pixel into fractions of classes, each representing a membership value (Ibrahim, Arora, and Ghosh, 2005).

There are four classification logics with respect to training of model in general: supervised, unsupervised, hybrid, and reinforced classification. So far, a variety of classification algorithms have been developed based on the two robust classification notions of 'hard' and 'soft'. All pixels in a particular satellite image indicate all various classes existing on the ground, rather than just one. However, there is the possibility of mixed classes within a pixel. Traditional hard classification is unable to generate appropriate classification results in such a case. When only a low spatial-resolution image is available or a regional scale-categorization is used, the possibilities of mixed pixels becomes more in images (Shalan, Arora, and Ghosh, 2003). Furthermore, because a fine spatial-resolution image contains more total pixels for a given geographic area, the number of mixed pixels decreases. There are different kinds of soft classifiers, such as conventional Maximum Likelihood Classifier (MLC) (Foody et al., 2000), Linear Mixture Model (LMM) (Sanjeevi and Barnsley, 2000; Lu and Weng, 2007), and theoretical fuzzy-set classifiers, commonly developed classifiers such as FCM (Fuzzy *c*-means) clustering (Bezdek, Ehrlich, and Full, 1984), PCM (possibilistic *c*-means) (Krishnapuram and Keller, 1993) clustering, modified versions of possibilistic classifiers, algorithms based on support-vector machines (SVM) (Vapnik, 1995), as well as the neural networks and their advanced versions (Li and Eastman, 2006).

In the fuzzy logic based classification, the diverse and vague nature of the physical world is considered, which can be applied in both supervised and unsupervised classification algorithms (Jensen, 1996). The process of evaluation of the proportion of a class from the composite spectrum of a sub-pixel is called 'pixel unmixing', or 'fuzzy classification' (Sanjeevi and Barnsley, 2000). Fuzzy *c*-means classification technique decomposes a pixel into different membership values according to the existing classes in the digital image. The relation of a pixel to the class is defined by membership function, and the quantitative value determined using this is called 'membership value', which lies between zero and one. A pixel strongly belongs to a class, if its membership value is close to one. If the value is close to zero, on the other hand, the pixel does not or only weakly belongs to that class. The sum of membership values of all classes within a pixel is equal to one (Bezdek et el., 1984, 1981). However, some fuzzy classifiers do not follow sum of membership values of all classes within a pixel constraint to be one.

To reduce the shortcomings of FCM, a possibilistic *c*-means (PCM) classifier has been proposed (Krishnapuram and Keller, 1993). Hyper line constraint was introduced in the PCM classifier objective function to restrict the transfer of a pixel vector to a particular class. The understanding of hyper line constraint is that the "sum of all the class's membership values in a pixel should be equal to one" (Krishnapuram and Keller, 1993; Kumar, Ghosh, and Dhadhwal, 2006). Further, PCM's modified version, which is known as modified possibilistic *c*-means algorithm, is able to reduce the time required to find the membership parameters in each iteration cycle (Li, Huang, and Li, 2003). As the results of classification based on PCM algorithm depends on selection of membership parameters, it is modified as enhanced PCM, which is able to rectify the drawback (Xie et al., 2008).

While generating Land Use/Land Cover (LULC) information a single-date, multi-spectral image is sufficient to provide necessary information. It is generally preferred if it has a high radiometric resolution and a wide range of digital number-values, which makes it to capture a wide variety of hues around an item. Spectral overlap between neighbouring classes happens commonly when performing level 2 or level 3 classifications with the same class. This can be prevented by using temporal data. To generate temporal data, two or more date acquisitions from the same geographic area might be employed. The number of temporal dates will also depend on the ability of a sensor to incorporate the unique changes occurring in the class of interest. The changes incorporated through these temporal images creates a unique signature of a particular class in a time domain. Because optical or passive remote-sensing images are cloud sensitive, the number of times required for temporal optical sets is limited. This problem can be handled by using images from other optical sensors from neighbouring dates in the form of a dual/multi-sensor method. Furthermore, if images from other optical sensors are unavailable owing to cloud cover, a synthetic aperture radar (SAR) sensor can be utilised to fill the temporal gaps. When employing multi-sensor images to build the required temporal image, different sensor parameters, such as spatial resolution, spectral resolution, indices parameters, back-scatter parameters, and so on should be used properly. Overall, it can be concluded that, when spectral information is insufficient to provide unique signature information about a class (classes that change over time, such as crops, vegetation, and post-disaster studies), temporal information from a mono/dual/multi-sensor can be used to provide very unique signature information about these classes.

With the increasing demand for users' time, there are a growing number of applications that necessitate a detailed assessment of a specific or single class. Crop-acreage estimation, specialised crop-mapping, inputs for individual crop-output estimation, deceased crop-fields mapping, and damaged crop-fields for crop insurance claims are some of the more prevalent examples. The single or specific class-mapping application can be for specific forest species, crop species, burnt paddy fields, and post-disaster damage-mapping, etc. While focusing on the crop studies, remote sensing data along with machine-learning models can provide solutions like specific-crop mapping, crop-acreage estimation, specific crop and its phenology, such as sowing, planting, and harvesting-stage monitoring, as well as crop insurance. Crop cycle may be monitored and mapped on an annual basis in a specific area utilising temporal single/dual/multi-sensor remote-sensing data and machine-learning algorithms.

All of the aforementioned objectives must be met, as concerns such as mixed pixels and noisy pixels can now be easily handled with remote-sensing raster images. While applying fuzzy machine-learning models, the mixed-pixel problem can be resolved. However, noisy pixels or isolated pixels can be addressed utilising particular fuzzy machine-learning models, as well as Markov random-field (MRF) or local-convolution approaches. Non-linearity across classes can also be managed using kernels in fuzzy machine learning models. Specific classes can be mapped using non-probability fuzzy models; spectral overlap between classes can be addressed using temporal images; and heterogeneity within a class can be handled by treating individual training samples as mean concepts. Furthermore, the advantages of using fuzzy machine-learning models are that these models require very small training-data sets.

1.2 INTRODUCTION TO MULTISPECTRAL REMOTE-SENSING

Multispectral remote-sensing is concerned with acquiring information through a spectral window ranging from visible to infrared in the electromagnetic consisting of multiple wide-wavelength bands. Since, at different wavelength bands, each material has its unique response in terms of reflection and absorption of electromagnetic radiations. Therefore, it is possible to differentiate among the different materials by observing their spectral response or spectral signatures in different remote-sensing images. Blue, green, yellow and red red-edge of visible bands, as well as near-infrared bands and short-wave infrared are typical examples of these large number of wavelength bands. Landsat, Quickbird, Sentinel-2, PlanetScope, Wordview-2/3, Spot, etc., are well-known satellite sensors—those are used as multispectral sensors in the present scenario.

To generate images, multispectral cameras employ many wavelengths of electromagnetic radiation. These cameras produce a series of images for that specific wavelength range. RGB colour image is a well-known multispectral image which consists of each of the red, green, and blue bands of EMR to sense collectively. These images are most commonly used for remote sensing applications. Multispectral images are captured from visible to short-wave, infrared spectral range in the electromagnetic spectrum—typically 2 to 15 spectral bands. Satellite imagery acquired using the different bands of the electromagnetic spectrum provides more than just a plain picture. Some sensors are able to detect energy beyond the visible range of the human eye and allow observation across the broad range of EMR.

Remote sensing is the process of capturing images of objects from an area by recording their reflected or emitted radiation in the electromagnetic spectrum recorded from a distance (typically from a satellite or aircraft). Special cameras collect the energy in the form of the remotely sensed images, which further help to identify information about the objects present in the area of interest. The ability or detection power of the sensor is measured in the form of four types of resolution of the sensor, viz., spatial, spectral, radiometric, and temporal. 'Spatial' regards the sensor's ability to discriminate between objects with reference to size. 'Spectral' regards the wavelength range; whereas 'radiometric' regards the number of available shades between extreme black and white. 'Temporal resolution' regards the revisit/repetivity frequency of the satellite sensor for the same geographical area.

1.3 INTRODUCTION TO HYPERSPECTRAL REMOTE-SENSING

Hyperspectral remote-sensing can be defined as acquiring images through a huge number of narrow, *contiguous* spectral bands from optical *spectral* range, while capturing spectra of different objects with very high-detail information. Within various remote sensing types, hyperspectral remote-sensing has lots of research opportunities (Chen and Qian, 2008). The hyperspectral sensors, whether they are airborne or spaceborne or ground-based, acquire images having narrow, contiguous spectral bands from visible range up to thermal IR range of the electromagnetic spectrum (Lillesand, Kiefer, and Chipman, 2004; Govender, Chetty, and Bulcock, 2007). To better define hyperspectral remote-sensing, some researchers have also called it 'imaging spectroscopy' (Mather and Koch, 2011). Overall, the main advantage of acquiring images from imaging spectroscopy is that it provides very precise spectral information about objects through narrow, contiguous spectral bands. These narrow, contiguous bands provide very specific properties of the objects from hyperspectral images. AVIRIS, a hyperspectral sensor, acquires the images in 224 contiguous bands of a geographical area, while multispectral sensors like TM sensor have just 6 bands. Thus, the resultant spectral curve for an object using AVIRIS imaging will be much more continuous, with more details.

In comparison to the multispectral, a hyperspectral sensor enables the identification of finer details of spectral variation of targets (Zha, Gao, and Ni, 2003). However, because the spectral dimension has risen, these strong spectral variations or characteristics come at the expense of classification accuracy, as well as high processing requirements. In hyperspectral remote-sensing, this leads to the dimensionality problem. The problem stems from the fact that the obtained hyperspectral data contains several spectral bands, yet the colour-composite image can only accommodate three—i.e. red, green, and blue (the three primary colours) (Mather and Koch, 2011). While extracting information from hyperspectral images, a great number of approaches have been devised to reduce the high dimensionality. Amongst them, Principal Component Analysis (PCA) (Silva, Centeno, and Henriques, 2010) and Minimum Noise Fraction (MNF) (Green et al., 1988) are common.

The classification of raw, hyperspectral bands is affected by the dimensionality problem—also termed the Hughes Phenomenon (Green, Berman, Switzer, and Craig, 1988). Hughes described this phenomenon as a relation between accuracy with dimensionality of images while keeping training data fixed (Hughes, 1968). It has been discovered that, as the dimensionality of the data increases while keeping constant training samples, the accuracy initially increases before declining after reaching its peak value. Also, while decreasing dimensionality of hyperspectral images, computational time reduces (Zha et al., 2003). In the literature, it can be found out that a large number of methods have been developed for reducing hyperspectral data dimensionality (Chen and Qian, 2007; Zha et al., 2003; Harsanyi and Chang, 1994; Hughes, 1968). Although indices have been applied for extraction of various properties of vegetation (Van Campenhout, 1978); Plaza, Martínez, Perez, and Plaza (2002); Zarco-Tejada, Miller, Morales, Berjón, and Agüera (2004); Haboudane, Miller, Pattey, Zarco-Tejada, and Strachan (2004), still there are a large number of indices developed for various other classes, which opens avenues for research.

Vane and Goetz (1988) recognised that imaging spectroscopy is a novel approach in the field of Earth remote-sensing. Vane and Goetz (1988) used Airborne Imaging Spectrometer (AIS) to discuss the capabilities of imaging spectroscopy. Sidjak and Wheate (1996) gathered photos from AIS for the first time in 1982 and raised concerns regarding high-dimensional data management and information extraction from AIS data sets. The topic of suitable methods to handle AIS data as well as processing software was highlighted in the context of AIS data management. Imaging spectroscopy, which produces images with around 200 consecutive bands, has been identified as highly useful data for recognising ground material and diagnosing small spectral characteristics (Thenkabail et al., 2013). With the development of imaging spectroscopy in the domain of Earth-observation imaging, there came presently developed sensors like AVIRIS (Estes, Hajic, and Tinney, 1983), Compact Airborne Spectrographic Imager (CASI) (Haralick, Shanmugam, and Dinstein, 1973), Digital Airborne Imaging Spectrometer (DIAS) (Clausi, 2002), and Airborne Prism EXperiment (APEX) Herold, Liu, and Clarke (2003), Datt, McVicar, Van Niel, Jupp, and Pearlman (2003).

A comparative review on multi-spectral versus hyperspectral remote-sensing was studied by Govender et al. (2007) while concentrating on the application of hyperspectral images in the areas of water resources and vegetation research. Further, a comparative literature review with respect to applications of hyperspectral images for the study of various resources discussed issues like flood-affected areas and flood monitoring, the quality of water in different water bodies, mapping of wetland areas, etc. The main strength of hyperspectral images—also known as imaging spectrometers—is towards imaging capability with contiguous bands from the visible spectrum to the shortwave, infrared region. Spectroscopy data has also been tested as an application for coral reef organism interaction study. This research study was able to demonstrate the capabilities of hyperspectral images in extracting interaction zone modifications and proving the presence of organisms in this zone, which information cannot be extracted using multispectral images (Thenkabail et al., 2013).

With further advancement in sensor imaging technology and the development of space based platform in addtion to the traditional airborne platform, more such research work have been done. Spaceborne hyperion data has been used for mineral mapping (Rodarmel and Shan, 2002). A comparative study using hyper-spectral data of airborne AVIRIS and spaceborne Hyperion has been done for mineral mapping (Kruse, Boardman, and Huntington 2003). A detailed methodology, including preprocessing steps as well as spatial mapping, while combining contiguous spectral and spectrally contiguous images through endmember spectra have also been discussed (Sidjak and Wheate (1999). It has also been tried to reduce image dimensions spatially and spectrally while extracting few key spectra of specific classes from hyperspectral images. Overall, the endmember-extraction approach facilitates the integration of spectral as well as spatial information from hyperspectral images. It has been concluded that approximately 60% of pixels were unlabelled from Hyperion images, in comparison to AVIRIS images for specific mineral mapping and hence the later image is better for mineral-mapping purposes. This study has proven the capability of airborne, hyperspectral images for specific mineral-class mapping.

Similarly, Kratt, Calvin, and Coolbaugh (2010) explored the mapping of geothermal indicator minerals while using hyperspectral remote-sensing images. Herold and Roberts (2005) integrated ground-based data with imaging spectroscopy for asphalt road-condition assessment. Aging as well as deterioration was seen while there was a transition from hydrocarbon roads to mineral-type roads. This study shows spectral evidence—an achievement towards the application of hyperspectral remote-sensing.

Imaging spectroscopy providing continuous spectra gives an opportunity for analysing biophysical properties of vegetation. Zhang, Chen, Miller, and Noland (2008) proposed a method for leaf chlorophyll content extraction using airborne, hyperspectral images. By this method, extensive field data was gathered and laboratory experimentation was done to get PROSPECT's model input parameters (Vane and Goetz, 1998). PROSPECT is a modified version of a generalized 'Plate model' developed by Allen et al. (1969; 1970). A hybrid model was developed while combining a geometrical, optical 4-scale model with a modified version of PROSPECT's model for estimating chlorophyll content using CASI images (Haralick et al., 1973). This hybrid model has produced results with good concurrence between measured and inverted leaf reflectance.

Goetz, Vane, Solomon, and Rock (1985), applied hyperspectral remote-sensing images for mapping invasive species due to disturbances in the river bed ecosystem in Japan. Logistic-regression method in binary mode was applied for invasive species mapping. On the logistic-regression model, original reflectance bands and Minimum Noise Fraction (MNF), as two variables, were applied for mapping Eragrostiscurvula (lovegrass) as an invasive species at the Ḳinu riverbed. From this study, it has been concluded that dimensionality reduction through MNF improves classification accuracy of invasive-species mapping. This research also draws conclusions about the strength of hyperspectral imagery to differentiate within invasive species and other crops.

Vane and Goetz (1988) concluded that phenological/plant stage information shows improvement in the accuracy of invasive-species mapping and gives much better spatial distribution of plant growth. Knox et al. (2013) proposed a phenological index (PhIX) and investigated its applicability along with the five existing vegetation indices for phenological studies. It was found that PhIX produced the best phenological classification accuracy. Although it has discriminated the seedling, dormant and matured phenological stages, however the flowering stage was found to be problematic. Overall, PhIX has been identified to be an important index for mapping the spatial deviation as well as supervising plant stages in savannah/grassland ecosystems (Vane and Goetz, 1988).

Nakazawa et al. (2012) mapped illegal poppy crop fields while separating with other crops using hyperspectral images. The objective was to compare these with previous, conventional change-detection techniques using IKONOS multi-spectral images. Nakazawa et al. (2012) used the spectral dissimilarity between poppy fields and wheat crops to be considered as an input parameter. In their study, the linear discriminant analysis using Red-Edge Position (REP) had produced quite efficient results to distinguish between poppy and wheat fields. Further, the partial, least-square discriminant analysis (PLSDA) was shown to be appropriate for the site, which was bordered by other crop fields.

Studying vegetation stress through hyperspectral remote-sensing is an important application. Chang, Westfield, Lehmann, Oertel, and Richter (1993) studied the stress seen in ash which was infected by emerald ash borer. Furthermore, the use of SpecTIR and VNIR spectrometers for mapping infected ash was investigated.

1.3.1 Hyperspectral Data Pre-processing

A large number of factors present during image acquisition affect hyperspectral data captured in very narrow, contiguous bands. Surface material reflects incident radiation from solar radiation incident on its surface in remote-sensing data acquisition. A material's spectral property is represented by the amount of reflected energy from its surface. So, surface reflectance is the portion of incident radiation that is reflected from the material's surface. Further, top of atmosphere (TOA) radiance is primarily measured by a sensor located at the top of the atmosphere. TOA radiance should be equal to material surface-reflectance under ideal conditions. Because atmospheric phenomena such as scattering and absorption affect incident and reflected radiation, TOA radiance is essentially a combination of surface reflectance, cloud reflectance, and scattering components based on air molecules and aerosol particles. Sensor gain and bias—also known as offset for each spectral band—influence the radiation values received by sensors, while raw data collected by a hyperspectral sensor is referred to as digital numbers (DNs).

Calibration of data is required for TOA radiance values in quantitative studies, and DNs values are later converted to surface reflectance.

The radiometric calibration process is used to convert DNs values to TOA radiance values. The process of removing atmospheric effects and estimating surface-reflectance values is known as atmospheric correction. Thus, the main corrections used in pre-processing are radiometric and atmospheric corrections.

The first step in radiometric calibration is to convert DN values to TOA radiance values. As a result, in order to calculate TOA radiance values, the sensor is calibrated for gain and bias in each spectral band as in eq. (1.1):

$$\text{Radiance}\ (L_\lambda) = (DN \times G_\lambda) + B_\lambda \tag{1.1}$$

where G_λ is gain value and B_λ is bias value for each spectral band (λ). The header of a hyperspectral image contains values for G_λ gain and B_λ bias.

The following step is to compute TOA reflectance values based on TOA radiance values. TOA reflectance is the ratio of TOA radiance to incident surface radiation. The following is the mathematical expression for calculating TOA reflectance (eq. 1.2):

$$\text{Reflectance}\ (\rho_\lambda) = \pi d^2\ L_\lambda\ ES_\lambda\ \alpha_E \tag{1.2}$$

where d is the Earth-sun distance in astronomical units, ES_λ is the mean solar irradiance for each spectral band, and α_E is the sun-elevation angle.

The second step in hyperspectral image pre-processing is atmospheric correction. TOA radiance/reflectance values are used to estimate surface-reflectance values in this application. Atmospheric-correction methods are classified into two categories. These are divided into two categories: empirical and model-based methods. Empirical methods are used to estimate relative surface-reflectance values based on a scene. Empirical methods are very computationally efficient and do not require any a priori measurements. Model-based methods, on the other hand, require in-situ atmospheric data and produce accurate estimates of surface-reflectance values. Some of the atmospheric-correction models developed are IARR, Flat field, Subtract dark pixel, RRS, Log residuals, Correctoob, Empirical line, Fast in scene, and SHARC.

Internal average relative reflectance (IARR) is an empirical approach that calculates relative surface-reflectance while using mean spectrum to normalise each pixel spectrum. The surface is assumed to be heterogeneous, and the spectral reflectance appearances are terminated. As a result, the mean spectrum from the surface resembles a flat-field spectrum. The IARR method is useful for calculating relative surface-reflectance values in areas devoid of vegetation. The flat-field correction method requires that the area being imaged be bright, uniform, and have neutral spectral-reflectance information. A mean spectrum of an area is calculated, which includes atmospheric scattering as well as the combined effects of solar irradiance and absorption. Relative surface reflectance values are calculated by dividing the pixel spectrum by the mean spectrum. Dark-pixel subtraction, also known as dark-object subtraction, is an empirical method that removes atmospheric haze from hyperspectral images much more effectively. The effect of atmospheric haze on the image is represented by high DN values, which make the image appear unnaturally bright. Dark pixels, on the other hand, are defined as having the lowest pixel values in each band. Dark pixels should have zero surface reflectance in general. Dark pixel values will be considered as an additive effect due to atmospheric path radiance. Remote sensing reflectance (RRS) primarily corrects atmospheric data with water bodies in hyperspectral ocean-colour data. The water-leaving radiance is calculated using this method.

When dividing hyperspectral data pixels by spectral and spatial, geometric means, logarithmic residual correction is used. The logarithmic residual-correction method is empirical in nature and requires statistics from the acquired hyperspectral image. The logarithmic residual-correction method is useful for removing the effects of solar irradiance and atmospheric transmittance. The out-of-band correction method, as the name suggests, removes out-of-band (OOB) effects from multispectral images. This method combines sensor spectral-response values with measured radiance. The empirical line-calibration method is based on the linear relationship between measured and surface reflectance. This method requires one or more target pixels from the input hyperspectral image, for which surface reflectance information is available. The empirical line-calibration approach necessitates statistics between measured spectral information and a priori target surface-reflectance information. When data is captured under uniform conditions but measurement related to target is time-variant, the empirical line-calibration approach can be used. When using in-scene characteristics, the fast in-scene method is applicable for atmospheric correction. An empirical method is the fast in-scene method. Correction parameters are computed using pixel spectra from the acquired hyperspectral image. Because the fast in-scene method uses approximate correction, it is computationally faster than model-based methods. To correct atmospheric effects, a fast in-scene method considers diverse pixel spectra and a sufficient number of dark pixels. This method computes the baseline spectrum from dark pixels. Using analytical solutions to the radiative transfer equation, the satellite hypercube atmospheric rapid-correction (SHARC) approach calculates absolute surface-reflectance information. Surface-reflectance values are calculated using atmospheric effects and the neighbour effect on the surface. This model can be used to calculate accurate surface-reflectance values while atmospheric-model parameters are available.

1.3.2 Endmember Extraction

In comparison to multi-spectral images, hyperspectral remote-sensing has far more applications, such as object detection, object identification, and discrimination of very specific surface materials. Some major pre-processing steps for hyperspectral images include atmospheric and radiometric corrections, data-dimensionality reduction, endmember extraction, and classification for spectral unmixing, as well as subpixel classification, if necessary. Image-based endmembers with the highest purity should be generated to achieve very high classification accuracy from hyperspectral data. Endmember in the hyperspectral domain is the purest, most spectrally distinct signature of a material. Many researchers have developed a large number of methods and algorithms for generating hyperspectral imagery-based endmembers. The majority of developed techniques rely on user-defined input parameters, but there are no specific standards for defining these parameters. Endmember extraction is inconsistent due to variations in defining parameters. There are numerous endmember-extraction algorithms proposed by various researchers, each with its own set of challenges and benefits, which have been discussed in the following analysis.

Endmember extraction methods are classified into two types. First, the category is based on the concept of convex geometry. Endmember extraction methods for convex geometry are divided into two categories: Orthogonal Projection (OP) and Simplex Volume (SV). Pixel Purity Index (PPI) (Boardman, Kruse, and Green, 1995), vertex component analysis (VCA) (Nascimento and Dias, 2003), and Sequential Maximum Angle Convex Cone (SMACC) are some Orthogonal Projection-based algorithms (Gruninger, Ratkowski, and Hoke, 2004). Simplex Volume (SV) based EE algorithms include N-FINDR (Winter, 1999), Simplex Growing Algorithm (SGA) (Chang, Wu, Liu, and Ouyang, 2006), and Convex Cone Analysis (CCA) (Ifarraguerri and Chang, 1999). Endmember extraction methods are also statistically based, with various orders such as second-order, higher-order, and so on, as well as a Swarm Intelligence-based endmember-extraction method. Iterative Error Analysis (IEA) (Neville, 1999) and endmember-extraction method using sparse component analysis (EESCA) are two methods that fall under second-order statistics (Wu, Feng, Xu, and Zhang, 2018). Skewness and the Independent Component Analysis EE algorithm (ICA-EEA) are two higher-order statistics-categories methods (Chang, 2013). The Ant Colony Optimisation (ACO) (Zhang, Sun, Gao, and Yang, 2011a), Particle Swarm Optimisation (PSO) (Zhang, Sun, Gao, and Yang, 2011b), and Artificial Bee Colony (ABC) are more advanced, SI-based endmember-extraction methods (Sun, Yang, Zhang, Gao, and Gao, 2015).

1.4 INTRODUCTION TO SAR REMOTE-SENSING

A layman may wonder, 'What is SAR?' SAR is an abbreviation for synthetic aperture radar, which was secretly developed during World War II. The primary goal of developing the SAR technique was to detect objects using a rangefinder method. Microwave pulses are transmitted and the reflected energy is recorded in the rangefinder method. In this process, the time of transmitted and returned pules is recorded and used to calculate the distance between the object and the sensor, as well as the velocity of light. This process can be understood by standing in front of a deep valley and speaking 'Hey' from the top of it for a few seconds. In this process, it is clear that the wave took its time to travel down the valley and bounced back to us from its lowest point. Once we know the one-sided wave travel time and the velocity of light, we can estimate the depth of the valley using this process. SAR in any domain, including space-based SAR, follows the same principle. Initially, the SAR technique was intended to be developed as a warning system for unwanted objects approaching defence forces for attack. All light contains electromagnetic waves, whether or not they are visible to human eyes. The type of light can be defined using frequency and wavelength in the electromagnetic spectrum. Radar uses microwave energy and has a frequency and wavelength. In comparison to visible light, microwave has a wavelength of the order of centimeter. With the passage of time, SAR (Synthetic Aperture Radar) was combined with radar to produce images of a specific resolution. Because radar-image resolution is proportional to sensor wavelength and antenna length, the use of synthetic aperture—also known as moniker synthetic—to increase resolution effectively increases the antenna length in SAR. The sensor sends a pulse from a super-long antenna that travels to the ground and is reflected back at different angles. Because super-long antennas cannot fly in space, antennas send pulses in rapid succession while collecting reflected pulses from various angles and times. So, there is a small antenna that sends a large number of pulses while another antenna collects reflected pulses from various angles, as a matter of fact it is called as synthetic. As a result, SAR is an active sensor with its own source.

Overall, SAR is a tool for measuring range. SAR detects a variety of objects by working with a wide range of frequency/wavelength bands in the microwave portion. There are various frequency/wavelength bands, such as the L-band, S-band, C-band, and X-band. These bands, which focus on roughness or surface property, are also dependent on range characteristics measured by the SAR sensor. Higher resolution can be achieved in SAR images with higher frequency, but sensor penetration-capability is reduced. Because of its high frequency, the Ka-band in the microwave region can provide higher resolution, however, it is absorbed by water moisture. As a result of these issues, the Ka-band is not used in SAR sensors. L and UHF bands can penetrate vegetation and are useful for estimating biomass and soil moisture. SAR can capture images through clouds, smoke, and ash. SAR images can detect burned foliage and trees as well as unburned trees that are still standing. Later, with higher temporal resolution, a fire-line can be detected using temporal SAR data. Because water and oil have different densities, SAR can map an oil spill on water very well. Backscatter coefficients, radar vegetation indices, and other parameters can be generated from SAR images for use in dual or multi-sensor data sets.

1.5 DIMENSIONALITY REDUCTION

In comparison to multi-spectral remote-sensing, hyperspectral remote-sensing has proven to be the most adaptable and information-rich source. It has applications in a variety of disciplines, such as specialised-class mapping, and provides opportunities for study in a variety of application fields. Although there are numerous advantages to using hyperspectral data, the biggest disadvantage is that it is difficult to follow Hughes Phenomenon due to high dimensionality. Conventional classifiers and algorithms fail to process hyperspectral data despite having less training data. Another aspect of categorisation is addressing the real-world issue of land-cover heterogeneity, or mixed pixels, which refers to the presence of many classes inside a single pixel.

Hyperspectral data must be capable of classifying each pixel and its dimension should be reduced without any loss of information (Harsanyi and Chang, 1994). Conventional classifiers do not give good accuracy with the high dimensionality of hyperspectral data. Dimensionality reduction can be done using a linear transformation (Harsanyi and Chang, 1994), such as PCA (Daughtry, Walthall, Kim, Colstoun, and McMurtrey, 2000) or MNF (Hughes 1968). Green, Berman, Switzer, and Craig (1998) have also discussed about the issues faced by statistical pattern algorithms due to high dimensionality of data. Hughes phenomena mentioned that classification accuracy first increases with increases in the number of dimensionality of data while having constant training samples. With rising data complexity and constant training samples, accuracy begins to decline. Itten et al. (2008) conducted additional research, highlighting a discrepancy in the Hughes phenomenon. Several studies analysed Hughes' approach in relation to other models and criticised it for using partial comparisons. Hughes' methodology was criticised, and the curse of dimensionality remained a barrier for hyperspectral data.

One of the well-established and accepted dimensionality-reduction methods is called Principal Component Analysis (PCA). Rodarmel and Shan (2002) explained the dimensionality-reduction method in detail. There was an increase in classification accuracy when using a subset of data when using hyperspectral data according to Hughes phenomenon. However, as greater dimensional data was employed, classification accuracy declined while training data remained constant. While generating PCA components for dimensionality reduction, it was discovered that the first ten PCA components had the most information. After the first ten PCA components, only noise was present and no valuable information (Daughtry et al., 2000). While considering processing time taken for classification with PCA components, it has been observed that processing time and effort has been reduced with respect to better accuracy achieved. In PCA data, orthogonal projections are used to maximise variance using PCA components and produce uncorrelated PCA components (Hughes, 1968). The number of times while applying PCA on hyperspectral data may not produce uncorrelated PCA components (Jehle et al., 2010). This gap provides a chance to investigate more effective dimensionality-reduction models.

The PCA technique employs linear transformation to generate new, uncorrelated PCA components while lowering information component by component (Kruse et al., 2003). Many times, PCA-based, decreased dimensionality is referred to as inherent dimensionality, with less emphasis on specific spectral classes and signatures of classes. Kruse et al. (2003) aimed to reduce data dimensionality by identifying pixels as pure or mixed. Individual-feature vectors are transformed as orthogonal subspaces of undesirable signatures in this method. Unwanted signals were afterwards eliminated, and the remaining feature vectors were transformed into interesting signatures to maximise signal-to-noise ratio. The final result was a single-element image that gives classification based on the signature's interest.

Plaza, Martínez, Plaza, and Pérez (2005) applied filtering and classification approaches on hyperspectral data through a series of extensive morphological-transformations steps. Generalised morphological alterations were applied concurrently while permitting spectral and spatial information collected from datasets and evaluated on urban and agricultural categorisation challenges. This proposed method produced excellent results when extracting spectrally close features' independent spatial attributes.

Chen and Qian (2007) tested Locally Linear Embedding (LLE) for proposing an approach as nonlinear dimensionality-reduction. The benefit of this method is that it preserves local topological structures using Laplacian eigen maps while maintaining locality characteristics in the form of data-point distances (Herold and Roberts, 2005). The achievement from this study was that proposed method can identify a large number of endmembers in the form of PCA as well as LLE, while endmember location accuracy was better.

From the input feature, set optimum features can be selected through feature-selection algorithms. But the methods through which new features are created through input-feature transformation are called feature-extraction algorithms (Zhang et al., 2008). Feature-extraction algorithms, while reducing the dimensionality of data, helps in reducing the processing time of the classification process. Furthermore, with a finite sample size, this feature-extracted data delivers superior classification accuracy in some circumstances (Jacquemoud and Baret, 1990). The literature shows a clear distinction between feature selection and feature-extraction measures applied to high-dimensional data. A detailed literature review on all dimensionality-reduction methods has also been conducted in a research paper (Silva, Centeno, and Henriques, 2010). A genetic algorithm for the feature-selection process has also been proposed. (Zha et al., 2003). The working of the genetic algorithm is based on a search technique following the natural evolutionary process. This literature survey provides detail about feature selection and extraction approaches to further extracting indices using airborne hyperspectral data.

Jain and Zongker (1997) also explored various models for feature-subset selection. They have created a large feature-dataset while using four different texture models for classifying SAR data. Subset feature selection was applied to avoid the curse of dimensionality. Classification accuracy was improved while combing features from different texture models, in place of using a single model. This study shows the importance of dimensionality reduction as well as feature-subset selection models. This study also shows the importance of subset-feature selection with limited training-data sets. Kohavi and John (1997) proposed filter-based and wrapper-based models for subset-feature selection. These subset-feature selection strategies also resulted in considerable improvements in classification accuracy.

Feature selection and extraction of dimensionality-reduction approaches have applications beyond analysis of hyperspectral data. Khoshelham and Elberink (2012) highlighted the importance of the dimensionality-reduction technique in damaged-building roofs mapping from airborne laser scanning data applying a segmentation approach. Experimental results showed that classification results were better while using a simple classifier with small training samples, in comparison to input features.

As per Janecek, Gansterer, Demel, and Ecker (2008), using feature selection and extraction approaches such as wrappers, filters, and PCA algorithms, a comparative classification accuracy research was conducted. Filters, wrappers, and embedding approaches were presented as three feature-selection and extraction techniques in total. Even though these techniques were not used on spatial remote-sensing data, a full description of PCA and feature-selection strategies was presented.

1.6 SUMMARY

This chapter discussed the use of remote sensing images, ranging from multispectral to hyperspectral images. The literature survey has also discussed about the importance of different levels of class mapping starting from thematic to single class, while applying various classification algorithms. The introductory section tried to cover about the use of remote sensing images for thematic mapping and the capability as well as challenges of single land cover identification. This chapter has also touched upon the importance of using multi-temporal remote-sensing data with a dual- or multi-sensor approach. An introduction about the role of fuzzy classifier and their capabilities of mapping the single or specific class, with small number of training samples has also been discussed. Sections 1.1, 1.2 and 1.3 have briefly covered about the multi-sensor, hyper-spectral and SAR remote sensing images, sensors and their applications respectively. In section 1.4, there was a discussion about dimensionality reduction, mainly linked with hyperspectral remote-sensing images. Finally, the significance of data-dimensionally reduction using a literature survey was explored. In the literature data, the dimensionality approach has been applied on multi-spectral data. The importance of data dimensionality to the form of spectral dimensionality-reduction has also been covered in Chapter 1. The importance of pixel-based various sorts of indices have been then discussed in Chapter 2. This indices approach is linked with temporal-indices generation in Chapter 3.

BIBLIOGRAPHY

Allen, W. A., Gausman, H. W., Richardson, A. J., and Thomas, J. R. (1969). *Interaction of isotropic light with a compact plant leaf*, J. Opt. Soc. Am. 59(10):1376–1379.

Allen, W. A., Gausman, H. W., and Richardson, A. J. (1970). *Mean effective optical constants of cotton leaves*, J. Opt. Soc. Am. 60(4):542–547.

Bezdek, J. C. (1981). *Pattern Recognition with Fuzzy Objective Function Algorithms*. New York and London: Plenum Press. https://doi.org/10.1007/978-1-4757-0450-1.

Bezdek, J. C., Ehrlich, R., and Full, W. (1984). FCM: The fuzzy c-means clustering algorithm. *Computers & Geosciences*, 10(2–3), 191–203. https://doi.org/10.1016/0098-3004(84)90020-7.

Boardman, J.W., Kruse, F.A. and Green, R.O. (1995). Mapping Target Signatures via Partial Unmixing of AVIRIS Data. Summaries, Fifth JPL *Airborn Earth Science Workshop*, 23–26 January 1995, JPL Publication 95-1, 1, 23-26.

Boochs, F., Kupfer, G., Dockter, K., and Kuhbaüch, W. (1990). Shape of the red edge as vitality indicator for plants. *International Journal of Remote Sensing*, 11(10), 1741–1753. https://doi.org/10.1080/01431169008955127.

Chang, C. I. (2013). *Hyperspectral Data Processing: Algorithm Design and Analysis*. Hoboken, NJ: John Wiley & Sons.

Chang, C. I., Wu, C. C., Liu, W., and Ouyang, Y. C. (2006). A new growing method for simplex-based endmember extraction algorithm. *IEEE Transactions on Geoscience and Remote Sensing*, 44(10), 2804–2819.

Chang, S.-H., Westfield, M. J., Lehmann, F., Oertel, D., and Richter, R. (1993). Channel airborne imaging spectrometer. In *Imaging Spectrometry of the Terrestrial Environment*, pp. 164–172. Orlando: SPIE, 1937.

Chen, G., and Qian, S.-E. (2007). Dimensionality reduction of hyperspectral imagery. *Journal of Applied Remote Sensing*, 1(1), 013509. https://doi.org/10.1117/1.2723663.

Chen, G., and Qian, S.-E. (2008). Simultaneous dimensionality reduction and denoising of hyperspectral imagery using bivariate wavelet shrinking and principal component analysis. *Canadian Journal of Remote Sensing*, 34, 447–454.

Cihlar, J. (2000). Land cover mapping of large areas from satellites: Status and research priorities. *International Journal of Remote Sensing*, 21(6–7), 1093–1114. https://doi.org/10.1080/014311600210092.

Clausi, D. (2002). An analysis of co-occurrence texture statistics as a function of grey level quantization. *Canadian Journal of Remote Sensing*, 28(1). https://doi.org/10.5589/m02-004.

Datt, B., McVicar, T. R., Van Niel, T. G., Jupp, D. L. B., and Pearlman, J. S. (2003). Preprocessing EO-1 Hyperion hyperspectral data to support the application of agricultural indexes. *IEEE Transactions on Geoscience and Remote Sensing*, 41(6), 1246–1259. https://doi.org/10.1109/TGRS.2003.813206.

Daughtry, C., Walthall, C., Kim, M. S., Colstoun, E. B., and McMurtrey, J. E. (2000). Estimating corn leaf chlorophyll concentration from leaf and canopy reflectance. *Remote Sensing of Environment*, 74, 229–239. https://doi.org/10.1016/S0034-4257(00)00113-9.

Estes, J. E., Hajic, E. J., and Tinney, L. R. (1983). Fundamentals of image analysis: Analysis of visible and thermal infrared data. In *Manual of Remote Sensing*, 2nd ed., pp. 987–1124. Edited by R. N. Colwell. Falls Church, VA: American Society for Photogrammetry and Remote Sensing.

Foody, G. M. (2000). Estimation of sub-pixel land cover composition in the presence of untrained classes. *Computers and Geosciences*, 26(4), 469–478. https://doi.org/10.1016/S0098-3004(99)00125-9.

Goetz, A., Vane, G., Solomon, J., and Rock, B. (1985). Imaging spectrometry for earth remote sensing. *Science* (New York, NY), 228, 1147–1153. https://doi.org/10.1126/science.228.4704.1147.

Govender, M., Chetty, K., and Bulcock, H. (2007). A review of hyperspectral remote sensing and its application in vegetation and water resource studies. *Water S.A.*, 33. https://doi.org/10.4314/wsa.v33i2.49049.

Green, A. A., Berman, M., Switzer, P., and Craig, M. D. (1988). A transformation for ordering multispectral data in terms of image quality with implications for noise removal. *IEEE Transactions on Geoscience and Remote Sensing*, 26, 65–74. https://doi.org/10.1109/36.3001.

Groten, S. M. (1993). NDVI-crop monitoring and early yield assessment of burkinafaso. *International Journal of Remote Sensing*, 14(8), 1495–1515. https://doi.org/10.1080/01431169308953983.

Gruninger, J. H., Ratkowski, A. J., and Hoke, M. L. (2004). The sequential maximum angle convex cone (SMACC) endmember model. In: Algorithms and technologies for multispectral, hyperspectral, and ultraspectral imagery X. *International Society for Optics and Photonics*, 5425, 1–15.

Haboudane, D., Miller, J. R., Pattey, E., Zarco-Tejada, P. J., and Strachan, I. B. (2004). Hyperspectral vegetation indices and novel algorithms for predicting green LAI of crop canopies: Modeling and validation in the context of precision agriculture. *Remote Sensing of Environment*, 90(3), 337–352, ISSN 0034–4257. https://doi.org/10.1016/j.rse.2003.12.013.

Haralick, R.M., Shanmugam, K. and Denstein, I. (1973). Textural Features for Image Classification. *IEEE Transactions on Systems, Man, and Cybernetics*, SMC-3, 610–621. https://doi.org/10.1109/TSMC.1973.4309314

Harsanyi, J. C., and Chang, C.-I. (1994). Hyperspectral image classification and dimensionality reduction: An orthogonal subspace projection approach. *IEEE Transactions on Geoscience and Remote Sensing*, 32(4), 779–785. https://doi.org/10.1109/36.298007.

Herold, M., Liu, X., and Clarke, K. (2003). Spatial metrics and image texture for mapping urban land use. *Photogrammetric Engineering and Remote Sensing*, 69, 991–1001. https://doi.org/10.14358/PERS.69.9.991.

Herold, M., and Roberts, D. (2005). Spectral characteristics of asphalt road aging and deterioration: Implications for remote-sensing applications. *Applied Optics*, 44, 4327–4334. http://doi.org/10.1364/AO.44.004327.

Hughes, G.: On the mean accuracy of statistical pattern recognizers. *IEEE Transactions on Information Theory* 14(1), 55–63 (January 1968). https://doi.org/10.1109/TIT.1968.1054102

Ibrahim, M. A., Arora, M. K., and Ghosh, S. K. (2005). Estimating and accommodating uncertainty through the soft classification of remote sensing data. *International Journal of Remote Sensing*, 26(14), 2995–3007. https://doi.org/10.1080/01431160500057806.

Ifarraguerri, A., and Chang, C. I. (1999). Multispectral and hyperspectral image analysis with convex cones. *IEEE Transactions on Geoscience and Remote Sensing*, 37(2), 756–770.

Itten, K. I., Dell'Endice, F., Hueni, A., Kneubühler, M., Schläpfer, D., Odermatt, D., Seidel, F., Huber, S., Schopfer, J., Kellenberger, T., Bühler, Y., D'Odorico, P., Nieke, J., Alberti, E., and Meuleman, K. (2008). APEX—the hyperspectral ESA airborne prism experiment. *Sensors*, 8, 6235–6259. https://doi.org/10.3390/s8106235.

Jacquemoud, S., and Baret, F. (1990). PROSPECT: A model of leaf optical properties spectra. *Remote Sensing of Environment*, 34(2), 75–91, ISSN 0034–4257. https://doi.org/10.1016/0034-4257(90)90100-Z.

Jain, A., and Zongker, D. (1997). Feature selection: Evaluation, application, and small sample performance. *IEEE Transactions on Machine Intelligence*, 19, 153–158.

Janecek, A., Gansterer, W., Demel, M., and Ecker, G. (2008). On the relationship between feature selection and classification accuracy. *Journal of Machine Learning Research—Proceedings Track*, 4, 90–105.

Jehle, M., Hueni, A., Damm, A., D'Odorico, P., Weyermann, J., Kneubiihler, M., Schlapfer, D., Schaepman, M. E., and Meuleman, K. (2010). APEX-current status, performance and validation concept. *Sensors*, 533–537.

Jensen, J.R. (1996). *Introductory Digital Image Processing: A Remote Sensing Perspective.* 2nd Edition, Prentice Hall, Inc., Upper Saddle River, NJ.

Jensen, J. R. (2010/1983). Biophysical remote sensing review article biophysical remote sensing. *Annals of the Association of American Geographers*, 73(1), 111–132. https://doi.org/10.1111/j.1467-8306.1983.tb01399.x.

Khoshelham, K., and Elberink, S. O. (2012). Accuracy and resolution of kinect depth data for Indoor mapping applications. *Sensors*, 12(2), 1437–1454. https://doi.org/10.3390/s120201437.

Knox, N., Skidmore, A., van der Werff, H., Groen, T., de Boer, W. F., Prins, H., Kohi, E., and Peel, M. (2013). Differentiation of plant age in grasses using remote sensing. *International Journal of Applied Earth Observation and Geoinformation*, 24, 54–62. https://doi.org/10.1016/j.jag.2013.02.004.

Kohavi, R., and John, G. H. (1997). Wrappers for feature subset selection. *Artificial Intelligence*, 97(1–2), 273–324, ISSN 0004-3702. https://doi.org/10.1016/S0004-3702(97)00043-X.

Kratt, C., Calvin, W., and Coolbaugh, M. (2010). Mineral mapping in the Pyramid Lake basin: Hydrothermal alteration, chemical precipitates and geothermal energy potential. *Remote Sensing of Environment*, 114, 2297–2304. https://doi.org/10.1016/j.rse.2010.05.006.

Krishnapuram, R., and Keller, J. M. (1993). A possibilistic approach to clustering. *IEEE Transactions on Fuzzy Systems*, 1(2), 98–110. https://doi.org/10.1109/91.227387.

Kruse, F., Boardman, J., and Huntington, J. (2003). Comparison of airborne hyperspectral data and EO-1 hyperion for mineral mapping. *IEEE Transactions on Geoscience and Remote Sensing*, 41, 1388–1400. https://doi.org/10.1109/TGRS.2003.812908.

Kumar, A., Ghosh, S. K., and Dhadhwal, V. K. (2006). Sub-pixel land cover mapping: SMIC system. *ISPRS Int. Sym."Geospatial Databases for Sustainable Development"*, Goa, India.

Li, K., Huang, H. K., and Li, K. L. (2003). A modified PCM clustering algorithm. *International Conference on Machine Learning and Cybernetics*, 2(November), 1174–1179. https://doi.org/10.1109/icmlc.2003.1259663.

Li, Z., and Eastman, J. R. (2006). The nature and classification of unlabelled neurons in the use of Kohonen's self-organizing map for supervised classification. *Transactions in GIS*, 10(4), 599–613. https://doi.org/10.1111/j.1467-9671.2006.01014.x.

Lillesand, T., Kiefer, R., & Chipman, J. (2004). *Remote Sensing and Image Interpretation* (5th ed.) Publisher: Wiley India (delhi).ISBN: 9788126513352, 8126513357

Lu, D., and Weng, Q. (2007). A survey of image classification methods and techniques for improving classification performance. *International Journal of Remote Sensing*, 28(5), 823–870. https://doi.org/10.1080/01431160600746456.

Mather, Paul M., and Koch, M. (2011). *Computer Processing of Remotely-Sensed Images: An Introduction*, 4th ed. John Wiley & Sons, Ltd, Hoboken, New Jersey. Print ISBN: 9780470742389, Online ISBN:9780470666517, https://doi.org/10.1002/9780470666517.

Nakazawa, A., Kim, J.-H., Mitani, T., Odagawa, S., Takeda, T., Kobayashi, C., and Kashimura, O. (2012). A study on detecting the poppy field using hyperspectral remote sensing techniques. *International Geoscience and Remote Sensing Symposium (IGARSS)*, 4829–4832. https://doi.org/10.1109/IGARSS.2012.6352532.

Nascimento, J. M., and Dias, J. M. (2003). Vertex component analysis: A fast algorithm to extract endmembers spectra from hyperspectral data. In *Iberian Conference on Pattern Recognition and Image Analysis*, pp. 626–635. Berlin and Heidelberg: Springer.

Neville, R. (1999). Automatic endmember extraction from hyperspectral data for mineral exploration. In *International Airborne Remote Sensing Conference and Exhibition*, 4th/21st Canadian Symposium on Remote Sensing, Ottawa, Canada.

Plaza, A., Martínez, P. J., Perez, R., and Plaza, J. (2002). Spatial/spectral endmember extraction by multidimensional morphological operations. *IEEE Transactions on Geoscience and Remote Sensing*, 40, 2025–2041. https://doi.org/10.1109/TGRS.2002.802494.

Plaza, A., Martínez, P. J., Plaza, J., and Pérez, R. (2005). Dimensionality reduction and classification of hyperspectral image data using sequences of extended morphological transformations. *IEEE Transactions on Geoscience And Remote Sensing*, 43(3).

Rodarmel, C., and Shan, J. (2002). Principal component analysis for hyperspectral image classification. *Surveying and Land Information Science; Gaithersburg*, 62(2), 115.

Roy, P. S., Behera, M. D., Murthy, M. S. R., Roy, A., Singh, S., Kushwaha, S. P. S., et al. (2015). New vegetation type map of India prepared using satellite remote sensing: Comparison with global vegetation maps and utilities. *International Journal of Applied Earth Observation and Geoinformation*, 39, 142–159. https://doi.org/10.1016/j.jag.2015.03.003.

Sanjeevi, S., and Barnsley, M. J. (2000). Spectral unmixing of Compact Airborne Spectrographic Imager (CASI) data for quantifying sub-pixel proportions of biophysical parameters in a coastal dune system. *Journal of the Indian Society of Remote Sensing*, 28(2–3), 187–204. https://doi.org/10.1007/BF02989903.

Shalan, M. A., Arora, M. K., and Ghosh, S. K. (2003). An evaluation of fuzzy classifications from IRS 1C LISS III imagery: A case study. *International Journal of Remote Sensing*, 24(15), 3179–3186. https://doi.org/10.1080/0143116031000094791.

Sidjak, R., and Wheate, R. (1996). Glacier mapping and inventory of the illecillewaet river basin, BC Canada, Using Landsat-Tm and Digital Elevation Data. *European Space Agency (Special Publication) ESA SP*, 391, 47.

Silva, C., Centeno, J., and Henriques, M. (2010). Automatic road extraction in rural areas, based on the Radon transform using digital images. *Canadian Journal of Remote Sensing*, 36, 737–749. https://doi.org/10.5589/m11-006.

Sun, X., Yang, L., Zhang, B., Gao, L., and Gao, J. (2015). An endmember extraction method based on artificial bee colony algorithms for hyperspectral remote sensing images. *Remote Sensing*, 7(12), 16363–16383.

Thenkabail, P., Mariotto, I., Gumma, M., Middleton, E., Landis, D., and Huemmrich, K. (2013). Selection of hyperspectral narrowbands (HNBs) and composition of hyperspectral twoband vegetation indices (HVIs) for biophysical characterization and discrimination of crop types using field reflectance and hyperion/EO-1 data. *IEEE Journal of Selected Topics in Applied Earth Observations and Remote Sensing*, 6, 427–439. https://doi.org/10.1109/JSTARS.2013.2252601.

Van Campenhout, J. M. (1978). On the peaking of the Hughes mean recognition accuracy: The resolution of an apparent paradox. *IEEE Transactions on Systems, Man, and Cybernetics*, 8(5), 390–395.

Vane, G., and Goetz, A. F. H. (1988). Terrestrial imaging spectroscopy. *Remote Sensing of Environment*, 24(1), 1–29, ISSN 0034–4257. https://doi.org/10.1016/0034-4257(88)90003-X.

Vapnik, V. (1995). *The Nature of Statistical Learning Theory*. New York: Springer-Verlag.

Winter, M. E. (1999). N-FINDR: An algorithm for fast autonomous spectral end-member determination in hyperspectral data. In: Imaging spectrometry V. *International Society for Optics and Photonics*, 3753, 266–276.

Wu, K., Feng, X., Xu, H., and Zhang, Y. (2018). A novel endmember extraction method using sparse component analysis for hyperspectral remote sensing imagery. *IEEE Access*, 6, 75206–75215.

Xie, Y., Sha, Z., and Yu, M. (2008). Remote sensing imagery in vegetation mapping: A review. *Journal of Plant Ecology*, 1(1), 9–23. https://doi.org/10.1093/jpe/rtm005.

Zarco-Tejada, P. J., Miller, J. R., Morales, A., Berjón, A., and Agüera, J. (2004). Hyperspectral indices and model simulation for chlorophyll estimation in open-canopy tree crops. *Remote Sensing of Environment*, 90, 463–476.

Zha, Y., Gao, J., and Ni, S. (2003). Use of normalized difference built-up index in automatically mapping urban areas from TM imagery. *International Journal of Remote Sensing*, 24, 583–594. https://doi.org/10.1080/01431160304987.

Zhang, B., Sun, X., Gao, L., and Yang, L. (2011a). Endmember extraction of hyperspectral remote sensing images based on the ant colony optimization (ACO) algorithm. *IEEE Transactions on Geoscience and Remote Sensing*, 49(7), 2635–2646.

Zhang, B., Sun, X., Gao, L., and Yang, L. (2011b). Endmember extraction of hyperspectral remote sensing images based on the discrete particle swarm optimization algorithm. *IEEE Transactions on Geoscience and Remote Sensing*, 49(11), 4173–4176.

Zhang, J. S., and Leung, Y. W. (2004). Improved possibilistic*c*-means clustering algorithms. *IEEE Transactions on Fuzzy Systems*, 12(2), 209–217. https://doi.org/10.1109/TFUZZ.2004.825079.

Zhang, Y., Chen, J. M., Miller, J. R., and Noland, T. L. (2008). Leaf chlorophyll content retrieval from airborne hyperspectral remote sensing imagery. *Remote Sensing of Environment*, 112, 3234–3247.

Why are indices covered in this chapter, and what role do they play in pre-processing?

To comprehend the various types of indices that can be computed from remote sensing images.

There are spectral and temporal attributes when working with multi-spectral temporal data. However, classifiers only work with one attribute at a time. Indices can help reduce spectral dimension, but to consider other attribute such as time.

CBSI approach for band selection in indices has also been covered. Advantage of CBSI approach is that for user only class location is required, bands information not required.

Grow through life. . . .

Do not go through life. . . .

2 Evolution of Pixel-Based Spectral Indices

2.1 INTRODUCTION

Jensen assessed the capability of satellite technology for identifying the underlying biophysical variables in 1983. Later, a human-perspective interpretation of spectral signatures of various aspects via colour was introduced (Carter, Cibula, and Dell, 1996; Vogelmann, Rock, and Moss, 1993; Swatantran, Dubayah, Roberts, Hofton, and Blair, 2011). It has been done while extracting the spectral signatures those have the capability to uniquely discriminate an object or a feature. The spectral signatures provide quantitative information of the interaction of Electro-Magnetic Radiation (EMR) with materials at different wavelengths. Later research has examined the interpretation of the roughness and texture of the surface or features, stating the capabilities of a sensor to capture these in terms of frequency changes and tonal arrangement in the image (Vogelmann et al., 1993; Swatantran et al., 2011).

The evolution of spectral indices has begun with the development of Simple Ratio (SR) (Birth and McVey, 1968) indices. Jordan, in the year 1969, was able to demonstrate the application of SR between the infrared and red reflectance for the estimation of Leaf Area Index (LAI) (Liu, Liang, Kuo, and Lin, 2004). The evolution of the spectral-band ratio resulted from the highly unique interaction of vegetation green canopies or leaves with electromagnetic radiations in the visible and infrared spectrum areas. It is well known that the presence of chlorophyll pigment in plants absorbs electromagnetic radiations in blue and red spectral regions and therefore is responsible for the green colour feature of all leaves. On the other hand, the moist-leaf internal structure strongly scatters the electromagnetic radiations in the infrared regions, which ultimately results in high reflectance. Thus, this unique spectral behaviour of green canopies has been utilised to develop various vegetation indices, which ultimately have been used to assess vegetation conditions using a remote-sensing image.

The Normalized Difference Vegetation Index (NDVI) by Rouse, Haas, Schell, and Deering (1974) is considered to be the extremely researched, applicable, and accepted index for vegetation-related applications. In this index, the reflectance in the red and infrared regions of the EMR is used to emphasise the vegetation features present in an area. For this purpose, the first study conducted using Earth Resources Technology Satellite (ERTS)-based LANDSAT-1 multi-spectral scanner data was in search of generating a band-ratio parameter having a high correlation with above-ground biomass and moisture content of vegetation features (Cho, Sobhan, Skidmore, and Leeuw, 2008). The outcome of the comprehensive research studies conducted in the Great Plains of the USA concluded that NDVI can be formulated as:

NDVI = (ρBand7 − ρBand5)/(ρBand7 + ρBand5), Where ρ is the reflectance

DOI: 10.1201/9781003373216-2

This index is widely applicable for the mapping of vegetation-based classes by separating them from the soil brightness and has quantitative values between −1 and +1. A modified version of NDVI has been proposed to cope up with it's negative values (Deering et al., 1975). This new index was named as Transformed Vegetation Index (TVI) and formulated as TVI = $\sqrt{(NDVI+0.5)}$. However, still the problem of negative values had remained the same and, technically, there was no difference in the outputs produced by these two vegetation indices in the form of the vegetation detection. The TVI index has also found a wide range of applications in various vegetation studies (Koetz et al., 2007; Blair, Rabine, and Hofton, 1999). Similarly, other vegetation indices, such as Corrected Transformed Vegetation Index (CTVI) (Perry and Lautenschlager, 1984) and Thiams Transformed Vegetation Index (TTVI) (Thiam, 1997), were also attempted from NDVI. Further, a reverse of Simple Ratio (SR) resulted in another vegetation index in the form of the Ratio Vegetation Index (RVI) by Richardson and Weigand (1977). RVI has been found to have strength and weaknesses identical to TVI (Ghosh, 2013). Like NDVI, the Normalized RVI (NRVI) (Baret and guyot, 1991) was generated from RVI by simply dividing (RVI-1) by (RVI+1). This index was also able to reduce the topographic, illumination, and atmospheric effects; likewise with other slope-based indices (Jackson and QiA., Huete, 1991). Besides its well-known applications, the NDVI has also been used to map the urban regions as inverse features or by excluding the NDVI values for vegetation-based features of an urban study-area (Peñuelas, Gamon, Fredeen, Merino, and Field, 1994). Since its development, many other variations of NDVI formulation have been attempted (Müller, Stadler, and Brandl, 2010). The popular NDVI has its own pitfalls, as it is not able to overcome the influence of soil reflectance (Rouse, 1974).

Inspired by the application of NDVI for vegetation, several other indices are developed for non-vegetative areas such as water and snow, considering them as important natural resources. For measuring the liquid water molecules in vegetation canopies, the Normalized Difference Water Index (NDWI) was proposed. This index utilizes two near-infrared spectral bands centred approximately at 0.86µm and 1.24µm and formulated as NDWI = $(\rho_{0.86\mu m} - \rho_{1.24\mu m})/(\rho_{0.86\mu m} + \rho_{1.24\mu m})$, where ρ is the reflectance. Similarly, the Normalized Difference Snow Index (NDSI) was generated using the visible band reflectance centred at 0.55 µm and SWIR reflectance near to 1.640µm using the formula NDSI = $(\rho_{0.55\mu m} - \rho_{1.64\mu m})/(\rho_{0.55\mu m} + \rho_{1.64\mu m})$.

Thus, the research and further studies on ratio-based indices prompted the evolution of various other spectral indices for mapping different land covers. In the case of vegetation-based land covers, these indices are developed to have a high correlation with various parameters such as leaf area, percentage green-cover, and biomass (Rouse et al., 1974). In 1988, research by Huete (Rouse et al., 1974) attempted to develop an index that was expected to remove the limitations caused by various atmospheric effects and background soil. In this research, he identified a meaningful influence of soil on the outcome of this index, especially for medial levels of green-vegetation canopy cover. Here the soil-vegetation spectral behaviour is modelled using the graphical adjustment done for the near infrared and red spectral regions. This index is named the Soil Adjusted Vegetation Index (SAVI) and formulated as follows Eq. (2.1):

$$SAVI = ((\rho nir - \rho red)/(\rho nir + \rho red + L)) \times (1 + L) \qquad \text{Eq. (2.1)}$$

This index minimises the effects of soil-background class by incorporating a parameter or correction factor or soil adjustment factor L in its equation. This parameter demands prior information about the areas of interest in terms of vegetation density. According to Huete, its value L takes the values 1, 0.5, and 0.25 for vegetation densities of categories low, intermediate, and high, respectively. If the value of this factor is 0, then the SAVI will be identical to the NDVI. Some studies suggested only a single adjustment factor using L = 0.5 for reducing the soil background effects for all vegetation densities; however, still, this could not have been an ideal factor for such a purpose. Qi et al. (1994) suggested another way to incorporate this correction factor to have a better correction for soil background-brightness under different vegetation cover or densities and thus proposed a Modified Soil Adjusted Vegetation Index (MSAVI). Here, in this factor, the correction factor L is omitted, and hence the prior information about the study area or vegetation-cover densities is not necessary. This particular index is formulated as in Eq. (2.2):

$$MSAVI = 0.5\,(2\rho nir + 1 - SQRT\,((2\rho nir + 1)^2 - 8(\rho nir - \rho red))) \qquad \text{Eq. (2.2)}$$

Elvidge and Chen later researched the detailed uses of narrow band technology in 1995. They compared the ability of small spectral bands to broad spectral bands for calculating LAI and the percentage of green cover using AVHRR, TM, and MSS data.

Elvidge and Chen (1995) employed these indices, together with three different derivatives of the green-vegetation index over time with the evolution of various indices using the contrasting nature of absorption in red and reflectance in NIR spectral regions. Amongst them, the first-order derivative of the green-vegetation index using a local baseline was called 1DL_DVGI, while the first- and second-order derivative of the green-vegetation index using a zero baseline were called 1DZ_DVGI and 2DZ_DVGI, respectively. The indices development and their various applications concluded that the ratio-based index or NDVI or slope-based indices are always influenced by the background conditions, while the Perpendicular Vegetation Index (PVI) or distance-based indices such as SAVI do not. Moreover, using the narrow spectral-bands data formulation of indices produces slightly better accuracy. Background measures can also be minimised using the derivative-based indices using the chlorophyll red-edge amplitude.

In year 2000, Broge and Leblanc (Roujean and Breon, 1995) studied the prediction power as well as the sensitivity analysis of various indices developed up to that time with respect to green LAI and Canopy Chlorophyll Density (CCD). They studied indices SR, NDVI, PVI (Jain and Chandrasekaran, 1982), SAVI (Rouse et al., 1974), and TSAVI (Jain, Duin, and Mao, 2000), etc., and also proposed a triangular vegetation index (Deering et al. 1975). Here, in this index, the absorbed energy by pigments is a function of the relative difference of spectral reflectance at red and near-infrared bands, as well as maximum value at the green band (Roujean and Breon, 1995). The index SAVI has been found to be least affected by the reflectance due to background and is found to be the best predictor of LAI. SAVI/MSAVI types of indices are selected to suppress soil impression while enhancing crop area. For low-density vegetation, the RVI is found to be an ideal index for estimating the LAI and CCD, while, for medium range vegetation densities, the triangular vegetation index is best-suited.

Daughtry, Walthall, Kim, Colstoun, and McMurtrey (2000) proposed Modified Chlorophyll Absorption Reflectance Index (MCARI) for estimating the chlorophyll content in corn leaves. Chlorophyll Absorption Reflectance Index (CARI) (Kim, 1994), as an original index, was effectively used to reduce the effects of non-photosynthetic materials on the spectral computations of photo-synthetically active radiation. The depth of the absorption taken by MCARI was at 0.67µm in place of the points at 0.55µm and 0.7µm. Though it had been proposed to compute the chlorophyll variations, yet this study had resulted the LAI, chlorophyll, and chlorophyll-LAI interactions accounted for 60%, 27%, and 13% of the variations of MCARI, respectively (Zarco-Tejada, Miller, Morales, Berjón, and Agüera, 2004; Barnes, Balaguer, Manrique, Elvira, and Davison, 1992). The MCARI has formulation as Eq. (2.3):

$$MCARI= ((\rho_{700}-\rho_{670}) -0.2\ (\rho_{700}-\rho_{550})) \times (\rho_{700}/\rho_{670}) \qquad \text{Eq. (2.3)}$$

The LAI (Haboudane et al. 2004 & Tejada et al., 2004) computation and effects of variation of chlorophyll on the relationships between LAI and VI were studied (Haboudane et al. 2004). The purpose was to develop a new vegetation index capable of identifying the LAI accurately without being susceptible to the chlorophyll variations. The PROSPECT (Vane and Goetz, 1998) and SAILH (Khoshelham and Elberink, 2012) radiative-transfer models were used to model this vegetation index. The derived indices MTVI1, MTVI2, and MCARI2 are given as in Eq. (2.4) to (2.6):

$$MTVI1 = 1.2[1.2(\rho_{800}-\rho_{550}) -2.5\ (\rho_{670}-\rho_{550})] \qquad \text{Eq. (2.4)}$$

$$MCARI2 = 3.75[(\rho_{800}-\rho_{670}) -3.75(\rho_{800}-\rho_{550})/SQRT\ [(2\rho_{800}+1)^2- 6\rho_{800}-55SQRT(\rho_{670}) -0.5] \qquad \text{Eq. (2.5)}$$

$$MTVI2 = 1.5[1.2(\rho_{800}-\rho_{550}) -2.5(\rho_{670}-\rho_{550})]/SQRT\ [(2\rho_{800}+1)^2 - (6\rho_{800}-5SQRT(\rho_{670})) -0.5] \qquad \text{Eq. (2.6)}$$

Applicability of these derived indices were tested using the CASI hyper-spectral data of real world (Haboudane et al. 2004 & Tejada et al., 2004). They found that the MCARI2 and MTVI2 are the most robust indices for estimating the green LAI. The new algorithms, modelled using the PROSPECT-SAILH models, were then applied on CASI data and have produced very positive results. Zarco-Tejada et al. (2001) have done a study in the search for an optical, narrow spectral-band index to identify the chlorophyll content and have attempted to establish a link amongst the leaf reflectance, transmittance, and canopy data. This study was found to be another landmark for the applicability of indices for narrow or hyperspectral-based applications. Here, CASI data was used along with the SAILH (Khoshelham and Elberink, 2012) model and MCRM, coupled with the leaf radiative-transfer model of PROSPECT (Vane and Goetz, 1998). Out of the various indices formulated, results of this study show that the red-edge-based indices are best suited for the leaf-colour chart estimation at canopy level.

Zarco-Tejada et al. (2004) conducted studies using combined indices such as MCARI, Transformed Chlorophyll Absorption Reflectance Index (TCARI), and Optimized Soil-Adjusted Vegetation Index (OSAVI) and found justifiable results for Cab estimation in the open-canopy crops. Further, studies were conducted to detect chlorosis and to monitor the physiological condition of Vitis vinifera L. (vineyard), using the Cab estimations at canopy and leaf levels (Zarco-Tejada et al. 2005; Plaza, Martinez, Plaza, and Perez, 2005). Swain and Davis (1978) conducted an extensive study for the application of remote sensing in Viticulture. They highlighted the need for research in precision viticulture using this technology and identified that the excellent indicator for Cab estimations is narrow band-based indices, whereas performance using traditional indices such as NDVI is poor in this regard. The inversion of the PROSPECT model, along with the field data, revealed that TCARI and OSAVI ratio (TCARI/OSAVI) was found to be the most consistent for estimating the Cab (Plaza et al., 2005).

After development of NDVI in the year 1974 by Rouse et al., many other indices have been formulated and tested in various studies like NDVI, ARVI (Kaufman and Tanre, 1992), and AFRI (Liu et al., 2004); NDVI (Rouse et al., 1974), RVI (Richardson and Weigand, 1977), NDVI (Jackson and Huete, 1991), NPCI (Normalized Pigment Chlorophyll Ratio Index) (Abderrazak, Khurshid, Staenz, and Schwarz, 2006), and PVI (Chehbouni, Qi, Huete, Kerr, and Sorooshian, 1994); Composite reflectance (Todd and Hoffer, 1998), DMSO (Barnes et al., 1992), SVI (Roujean and Breon, 1995), DVI (Difference vegetation index) (Elvidge and Chen, 1995), and MCARI (Zarco-Tejada et al., 2004). Every index mentioned or discussed previously was developed to meet a specific objective or for a particular application. These indices are more explorable due to the availability of data acquired in narrow bands using hyperspectral sensors. Identification of properties such as chlorophyll content, leaf-area index, plant stress, etc. are only presumable owing to the capability of the acquisition of data in the fine spectral-width of these sensors.

In many studies, the estimation of biomass production has been conducted using the indices applied on hyperspectral data. Tucker (1979) attempted to develop a predictor that helped, on a yearly basis, with the monitoring of the production of the biomass of grass or herbs. They have considered the spectral indices and linear-regression model for this purpose and established a high correlation between the biomass and vegetation indices. Moreover, the prediction error for modelling using indices data was much less while utilising the red-edge band. A similar study conducted by Cho et al. (2008) (Haboudane, 2004) used the red-edge-based index as a possible method for discrimination of vegetation species using hyperspectral data at leaf and canopy levels.

Amongst the various indices-based researches, the study by Richardson and Weigand (1977) demonstrated the design of an optimal index using spectral reflectance at red and NIR channels of AVHRR satellite data for remote-sensing-based applications. This was to consider the perturbing effects and formulate the Global Environmental Monitoring Index (GEMI). This work has established another way to formulate indices to work in an optimal manner for various applications related to vegetation identification in a multispectral domain and thus has acted as a subset of the spectral indices (Richardson and Weigand, 1977).

2.2 SPATIAL INFORMATION

According to Huete (1998), the spatial indices can be interpreted in different ways. It is due to the fact that spatial information is generally interpreted as the distribution of the objects in an image, as well as the influence of neighbouring pixels. Defining the characteristic for obtaining information from an image is equivalent to defining its type that human beings used in interpretation (Carter et al., 1996). To do so, the most important elements can be of pattern-type elements such as spectral, textural, and contextual features (Carter et al., 1996). This way of analysis was supported by Estes, Hajic, and Tinney (1983) (Vogelmann et al., 1993) by considering the tonal/colour variations as primary interpretation-keys for visual interpretation. Later, in many studies, researchers have applied textural-based information to identify an object or a region of interest spatially (Carter et al., 1996).

Though many research studies were conducted for spatial information extraction before the Haralick, Shanmugam, and Dinstein (1973) study, still, none of them modelled or applied the texture-based study for this purpose. Combining textural information with spatial metrics (Gitelson and Merzlyak, 1997) has improved land-cover information-extraction accuracy from high-resolution satellite imagery. It is done by considering texture as a feature for the input-data set.

2.3 SPECTRAL INDICES

The spectral indices are extensively researched and have shown wide applications in emphasising the certain classes of interest while de-emphasising the other classes of non-interest. Amongst various spectral indices, the NDVI is most commonly used for extraction of vegetation-related classes (Lin, Qin, Dong, and Meng, 2012). This index was first implemented by Rouse et al. (1974) and is a numerical indicator using the NIR and red spectral-bands of electromagnetic radiation (Cho et al., 2008). It is generally more preferred for identifying green vegetation against the previously identified, simple ratio-based index (Liu et al., 2004). However, the main difficulty of using NDVI is due to its high correlation with the multiple vegetation-parameters, as well as the influence of soil-background reflectance (Rouse et al., 1974), which many times makes it difficult to separate various vegetation classes within themselves. Therefore, various spectral indices to identify the chlorophyll content (Jordan, 1969; Jackson and Huete, 1991; Todd and Hoffer, 1998), Leaf Area Index (LAI) (Barnes et al., 1992; Roujean and Breon, 1995; Zarco-Tejada et al., 2004), and water content (Gamon, Penuelas, and Field, 1992), were developed.

Except NDVI, many other indices for extraction of vegetation classes (Zarco-Tejada et al., 2004; Zarco-Tejada, Berjon, Lopezlozano, Miller, Martin, Cachorro, Gonzalez, and Defrutos, 2005) as well as for different types of land cover have also been formulated. Amongst the non vegetation index category, the utility of Normalized Difference Built-Up Index (NDBI) (Peñuelas et al., 1994; Zha, Gao, and Ni 2003) a Normalized Difference Water Index (NDWI) (Gamon et al., 1992; Gao 1996), and Normalized Difference Snow Index (NDSI) (Peñuelas et al.; 1994) are tested.

The definition of spectral indices depends on their capability of identifying the unique spectral curve or property/behaviour of a specific object. These are always used to emphasise a specific object or class. With technological advancement, multi-spectral remote-sensing is shifting towards the narrow band or hyperspectral remote-sensing having a greater number of spectral details, hence making the specific vegetation-class or object identification much easier (Haboudane, 2004; Lin et al., 2012; Cho et al., 2008).

Because of the enormous number of spectral indices produced so far for diverse vegetation or other land cover/classes, the selection of an ideal index is always dependent on the type of object or research site, the nature of the dataset, and other parameters. As a result, the advantage of applying the spectral-indices approach for extracting information from remote-sensing data is extremely important, and these methods are favoured over alternative transformations for this purpose.

2.4 TEXTURE-BASED SPATIAL INDICES

The spatial-based indices are defined in many ways; for example, in terms of geospatial indices, it was used as an Area Index, Shape Index, etc., for measuring the urban sprawl in Beijing (Baret, Guyot, and Major, 1989), while some other researchers used it to analyse them for segregation; for example, spatial distribution of population groups as a spatial element (Huete, 1998). Some other studies analysed the spatial metrics in combination with remote-sensing data for deriving and modelling the information of land use/land cover changes. On the other hand, texture provides the spatial distribution of tonal variations in a particular spectral band (Huete, 1998). It is well known that extensive information can be observed in remote-sensing imagery when it is visually analysed. This information takes into account the context, edge, texture, tonal, and colour variations Haralick et al. (1973) (Huete, 1998; Kim, 1994; Verhoef, 1984). According to Haralick et al. (1973) (Carter et al., 1996) the most important elements for identification of land cover or land uses from a remote-sensing image can be spectral, textural, and contextual features.

According to Estes et al. (1983) (Gitelson and Merzlyak, 1997), the colour or tone are primary or basic elements of a visual-image interpretation, followed by size, shape, and texture. All these are the spatial elements of the interpretation. Many research studies nowadays are incorporating them to match the human-based interpretation characteristics while performing a digital-image analysis or classification (Verhoef, 1984). The tone or colour in a digital image is due to the set of all pixels having the identical or nearly-identical brightness value (Verhoef, 1984). If there is a region where the pixels have wide tonal-variations or discrete tonal-features, then the property texture is more applicable (Verhoef, 1984). Jensen, in the year 1996, provided a basic understanding of textural-based analysis for interpreting the remote-sensing imagery.

Understanding the need to utilise texture analysis for identifying the objects or regions of interest in an image (Vogelmann et al., 1993), Haralick et al. (1973) studied its applicability for identifying some objects from remote-sensing data. Textural analysis and its validity have been tested on three types of remote-sensing-based data: photo-micrographs, panchromatic aerial-photographs of scale 1:20000, and ERTS MSS imagery. To derive information at the level of human perspective, this research work has highlighted the significance of spectral, spatial, and contextual information for identifying objects. Previously conducted studies were limited to developing algorithms for understanding coarseness and edges. In this study, all the textures were derived from angular, nearest-neighbour grey-tone spatial-dependence matrices or co-occurrence matrices (Vogelmann et al., 1993; Gitelson and Merzlyak, 1997) and considered the statistical nature of textures for analysis. Results of this study showed that the accuracies achieved were 89%, 82%, and 83% for object identification in photo-micrographs, aerial photographs, and MSS-image data, respectively.

Thus, textural information has been used by various studies for describing objects spatially in digital-image data. Whereas, spatial metrics have been widely used in an urban environment for depicting the spatial distribution of housing and populations (Gitelson and Merzlyak, 1997). Herold, Liu, and Clarke (2003) (Gitelson and Merzlyak, 1997) have explored a combination of texture information and spatial metrics to describe the urban spatial-characteristics qualitatively. In this study, an object-oriented approach had been used to classify large urban areas.

A Gray-Level Co-occurrence Matrix (GLCM) texture has been used to describe homogeneity. The GLCM texture was found to be most suitable to provide spatial information and achieved an overall classification accuracy of 76.20% for land-use classification. GLCM has also been used to interpret or classify the Synthetic Aperture Radar (SAR) imagery data (Zarco-Tejada, Miller, Noland, Mohammed, and Sampson, 2001; Clausi, 2002).

Here, in this study, the co-occurrence probabilities have provided a second-order method for generating texture features and hence for interpreting or classifying SAR sea-ice imagery. For classification of the SAR dataset, the Fisher Linear Discriminant (FLD) was applied. This study concluded that it is not wise to use all the texture-analysis features collectively; however, a few might add sufficient information for classification of the objects. Statistical parameters such as correlation, contrast, and entropy are the most suitable and independent parameter to be added as features.

2.5 SUMMARY

In this chapter, various spatial, spectral, texture-based indices have been discussed. The importance of these indices has been discussed here for single-class extraction. Next, Chapter 3 is related to the use of multi-sensor, temporal remote-sensing and its applications for single-class extraction.

BIBLIOGRAPHY

Abderrazak B., Khurshid S., Staenz K., and Schwarz J. (2006). Wheat crop chlorophyll content estimation from ground-based reflectance using chlorophyll indices, *IEEE International Symposium on Geoscience and Remote Sensing*, pp. 112–115. https://doi.org/10.1109/IGARSS.2006.34.

Baret F., Guyot G., and Major D. J. (1989). TSAVI: A vegetation index which minimizes soil brightness effects on LAI and APAR estimation, *Geoscience and Remote Sensing Symposium, 1989. IGARSS'89. 12th Canadian Symposium on Remote Sensing., 1989 International*, vol. 3, pp. 1355–1358.

Barnes J. D., Balaguer L., Manrique E., Elvira S., and Davison A. W. (1992). A reappraisal of the use of DMSO for the extraction and determination of chlorophylls a and b in lichens and higher plants, *Environmental and Experimental Botany*, vol. 32, no. 2, pp. 85–100.

Belkin M., and Niyogi P. (2003). Laplacian eigenmaps for dimensionality reduction and data representation, *Neural Computation*, vol. 15, no. 6, pp. 1373–1396.

Carter G. A., Cibula W. G., and Dell T. R. (1996). Spectral reflectance characteristics and digital imagery of a pine needle blight in the southeastern United States, *Canadian Journal of Forest Research*, vol. 26, no. 3, pp. 402–407.

Chehbouni, A., Qi, J., Huete, A. R., Kerr, Y. H., and Sorooshian, S. (1994). A modified soil adjusted vegetation index, *Remote Sensing of Environment*, vol. 48, no. 2, pp. 119–126.

Cho M. A., Sobhan I., Skidmore A. K., and Leeuw J. de (2008). Discriminating species using hyperspectral indices at leaf and canopy scales, in *Proceedings of ISPRS Congress*, Beijing, pp. 369–376.

Daughtry C. S. T., Walthall C. L., Kim M. S., Colstoun E. B. De, and McMurtrey J. E. (2000). Estimating corn leaf chlorophyll concentration from leaf and canopy reflectance, *Remote Sensing of Environment*, vol. 74, no. 2, pp. 229–239.

Elvidge C. D., and Chen Z. (1995). Comparison of broad-band and narrow-band red and near-infrared vegetation indices, *Remote Sensing of Environment*, vol. 54, no. 1, pp. 38–48.

Estes J. E., Hajic E. J., and Tinney L. R. (1983). Fundamentals of image analysis: Analysis of visible and thermal infrared data, *Manual of Remote Sensing*, vol. 1, pp. 987–1124.

Gamon J. A., Penuelas J., and Field C. B. (1992). A narrow-waveband spectral index that tracks diurnal changes in photosynthetic efficiency, *Remote Sensing of Environment*, vol. 41, no. 1, pp. 35–44.

Gao B. (1996). NDWI—A normalized difference water index for remote sensing of vegetation liquid water from space, *Remote Sensing of Environment*, vol. 58, no. 3, pp. 257–266.

Gitelson A. A., and Merzlyak M. N. (1997). Remote estimation of chlorophyll content in higher plant leaves, *International Journal of Remote Sensing*, vol. 18, no. 12, pp. 2691–2697.

Haboudane D. (2004). Hyperspectral vegetation indices and novel algorithms for predicting green LAI of crop canopies: Modeling and validation in the context of precision agriculture, *Remote Sensing of Environment*, vol. 90, no. 3, pp. 337–352.

Haralick R. M., Shanmugam K., and Dinstein I. (1973). Textural features for image classification, *IEEE Transactions on Systems, Man and Cybernetics*, vol. SMC-3, no. 6, pp. 610–621.

Herold M., Liu X., and Clarke K. C. (2003). Spatial metrics and image texture for mapping urban land use, *Photogrammetric Engineering and Remote Sensing*, vol. 69, no. 9, pp. 991–1002.

Huete A. R. (1998). A soil-adjusted vegetation index (SAVI), *Remote Sensing of Environment*, vol. 25, no. 3, pp. 295–309.

Jackson R. D., and Huete A. R. (1991). Interpreting vegetation indices, *Preventive Veterinary Medicine*, vol. 11, no. 3, pp. 185–200.

Jensen J. R. (1983). Biophysical remote sensing, *Annals of the Association of American Geographers*, vol. 73, no. 1, pp. 111–132.

Jordan C. F. (1969). Derivation of leaf-area index from quality of light on the forest floor, *Ecology*, vol. 50, no. 4, pp. 663–666.

Kaufman, Y. J., & Tanré, D. 1992. Atmospherically Resistant Vegetation Index (ARVI) for EOS-MODIS. *IEEE Transactions on Geoscience and Remote Sensing*, 30, 261–270.

Khoshelham K., and Elberink S. O. (2012). Role of dimensionality reduction in segment-based classification of damaged building roofs in airborne laser scanning data, in *Presented at the 4th GEOBIA*, Rio de Janeiro, Brazil.

Kim M. S. (1994). The use of narrow spectral bands for improving remote sensing estimation of fractionally absorbed photosynthetically active radiation (fAPAR), *Masters, Department of Geography*, University of Maryland, College Park, MD.

Koetz B., Sun G., Morsdorf F., Ranson K. J., Kneubühler M., Itten K., and Allgöwer B. (2007). Fusion of imaging spectrometer and LIDAR data over combined radiative transfer models for forest canopy characterization, *Remote Sensing of Environment*, vol. 106, no. 4, pp. 449–459.

Lin P., Qin Q., Dong H., and Meng Q. (2012). Hyperspectral vegetation indices for crop chlorophyll estimation: Assessment, modeling and validation. *IEEE International Geoscience and Remote Sensing Symposium, Munich, Germany, 2012*, pp. 4841–4844, doi: 10.1109/IGARSS.2012.6352529.

Liu G. R., Liang C. K., Kuo T. H., and Lin T. H. (2004). Comparison of the NDVI, ARVI and AFRI vegetation index, along with their relations with the AOD using SPOT 4 vegetation data, *Terrestrial, Atmospheric and Oceanic Sciences*, vol. 15, no. 1.

Müller J., Stadler J., and Brandl R. (2010). Composition versus physiognomy of vegetation as predictors of bird assemblages: The role of lidar, *Remote Sensing of Environment*, vol. 114, no. 3, pp. 490–495.

Peñuelas J., Gamon J. A., Fredeen A. L., Merino J., and Field C. B. (1994). Reflectance indices associated with physiological changes in nitrogen-and water-limited sunflower leaves, *Remote Sensing of Environment*, vol. 48, no. 2, pp. 135–146.

Plaza A., Martinez P., Plaza J., and Perez R. (2005). Dimensionality reduction and classification of hyperspectral image data using sequences of extended morphological transformations, *IEEE Transactions on Geoscience and Remote Sensing*, vol. 43, no. 3, pp. 466–479.

Richardson A. J., and Weigand C. L. (1977). Distinguishing vegetation from soil background information, *Photogrammetric Engineering and Remote Sensing*, vol. 43, no. 12, pp. 1541–1552.

Roujean J.-L., and Breon F.-M. (1995). Estimating PAR absorbed by vegetation from bidirectional reflectance measurements, *Remote Sensing of Environment*, vol. 51, no. 3, pp. 375–384.

Rouse, J.W., Haas, R.H., Schell, J.A. and Deering, D.W., 1974. Monitoring vegetation systems in the Great Plains with ERTS. *Third ERTS Symposium*, NASA SP-351, 309-317.

Swain P. H., and Davis S. M., Eds. (1978). *Remote Sensing: The Quantitative Approach*. London and New York: McGraw-Hill International Book Co.

Swatantran A., Dubayah R., Roberts D., Hofton M., and Blair J. B. (2011). Mapping biomass and stress in the Sierra Nevada using lidar and hyperspectral data fusion, *Remote Sensing of Environment*, vol. 115, no. 11, pp. 2917–2930.

Todd S. W., and Hoffer R. M. (1998) Responses of spectral indices to variations in vegetation cover and soil background, *Photogrammetric Engineering and Remote Sensing*, vol. 64, pp. 915–922.

Tucker C. J. (1979). Red and photographic infrared linear combinations for monitoring vegetation, *Remote Sensing of Environment*, vol. 8, no. 2, pp. 127–150.

Van Campenhout J. M. (1978). On the peaking of the Hughes mean recognition accuracy: The resolution of an apparent paradox, *IEEE Transactions on Systems, Man and Cybernetics*, vol. 8, no. 5, pp. 390–395.

Vane G., and Goetz A. F. (1998). Terrestrial imaging spectroscopy, *Remote Sensing of Environment*, vol. 24, no. 1, pp. 1–29.

Verhoef W. (1984). Light scattering by leaf layers with application to canopy reflectance modeling: The SAIL model, *Remote Sensing of Environment*, vol. 16, no. 2, pp. 125–141.

Vogelmann J. E., Rock B. N., and Moss D. M. (1993). Red edge spectral measurements from sugar maple leaves, *International Journal of Remote Sensing*, vol. 14, no. 8, pp. 1563–1575.

Zarco-Tejada P. J., Miller J. R., Noland T. L., Mohammed G. H., and Sampson P. H. (2001). Scaling-up and model inversion methods with narrowband optical indices for chlorophyll content estimation in closed forest canopies with hyperspectral data, *IEEE Transactions on Geoscience and Remote Sensing*, vol. 39, no. 7, pp. 1491–1507.

Zha Y., Gao J., and Ni S. (2003). Use of normalized difference built-up index in automatically mapping urban areas from TM imagery, *International Journal of Remote Sensing*, vol. 24, no. 3, pp. 583–594.

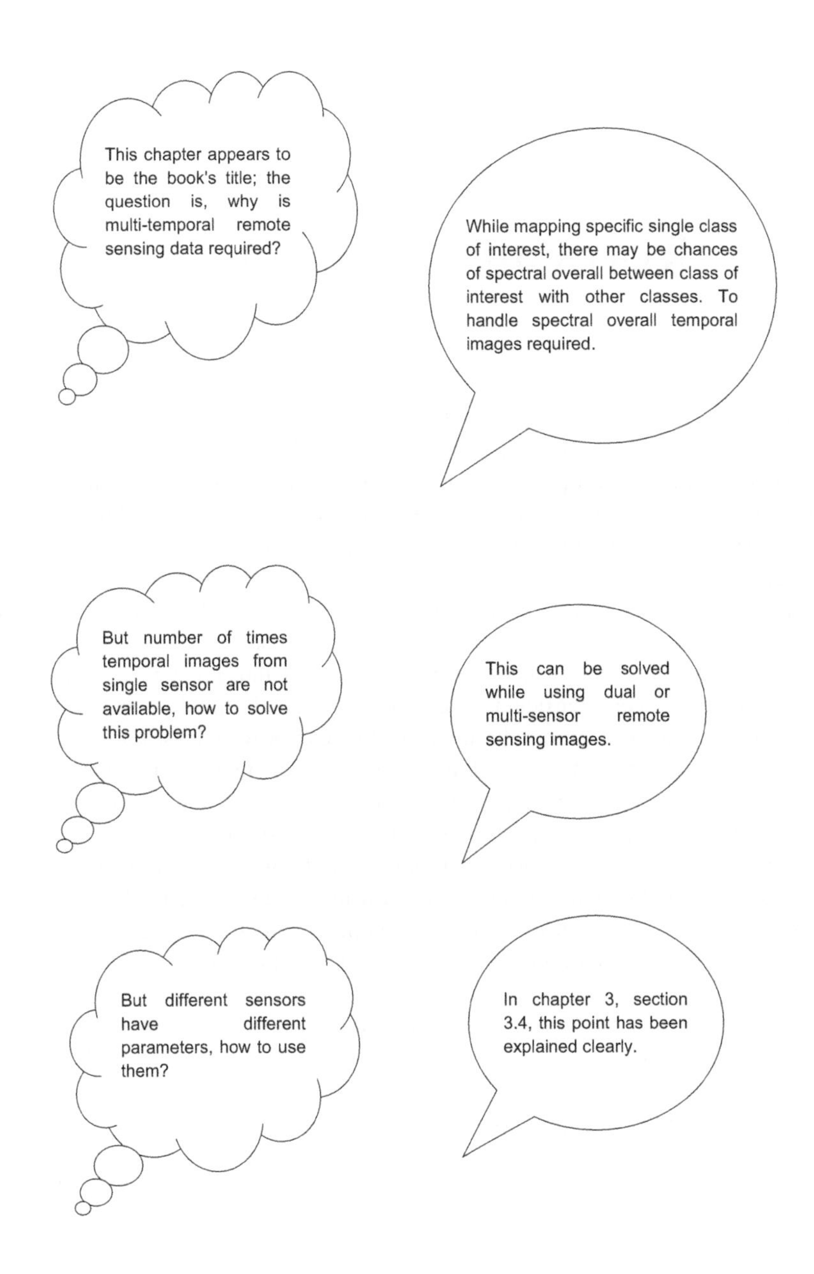

When everything is going against, remember airplane takes off against wind. . . .
What stop us to do anything in this life. . . .
Just two words — "What If?"

3 Multi-Sensor, Multi-Temporal Remote-Sensing

3.1 INTRODUCTION

The basic prerequisite for deducing outcomes from remote-sensing data is the generation of thematic maps. Image data from multispectral remote-sensing is essential for creating thematic maps. It provides thematic information at level-one classification. However, if thematic information is to be extracted within specific land-use/land-cover class as a level two or level three, there is a chance of spectral overlap with other classes. Spectral overlap between classes exists, when there are various classes or land cover of nearly identical spectral signatures. This can be understood with an example: suppose we are interested in extracting Rabi season crops like wheat and mustard separately; then, chances are that wheat and mustard will have the same spectral information on a particular date, especially in their initial stage of growth. To handle such a situation, temporal remote-sensing can provide least-correlated information in a time domain for extracting a single class of interest. Temporal remote-sensing data provides unique stages information of a specific crop of interest and hence produces a distinctive spectral-signature profile for that. Thus, it helps to map a specific class of interest, as well as produce unique information in forest-species mapping, crop stubble burning fields, post-disaster loss assessment, etc.

In this chapter, the processing steps of temporal remote-sensing are given. In a single quote, importance of temporal images can be expressed as:

> ***'when spectral information stops, temporal information starts to map a specific class of interest'.***

3.2 TEMPORAL VEGETATION INDICES

Temporal remote-sensing data-information is critical for understanding land-cover dynamics. Several research studies have already been undertaken using time series data for crop categorisation and identification of a specific or single class of interest.

Wardlow and Egbert (2008) used time-series data of MODIS normalised-difference vegetation index for mapping land-use/land-cover (LULC) classes over agricultural areas of US using a decision-tree classifier. They discovered that most crops can be separated at a specific point in their growing season. It was also shown that the spectral separability of a specific crop is influenced by intra-class variability caused by meteorological conditions, as well as the date of plantation. In their study, they discovered a high association between the Enhanced Vegetation Index (EVI) and the NDVI during the senescence period of the crop cycle, with a classification accuracy of more than 80%.

DOI: 10.1201/9781003373216-3

Crop phenology mapping was conducted by Sakamoto et al. (2005) from time series MODIS data. They used the EVI and were able to determine the phenological stages—viz., planting date, heading date, growing period and harvesting date—using the wavelet transformation. Serra et al. (2008) proposed a method for monitoring and mapping the temporal signatures of the crops and their inter-annual dynamics in the Mediterranean area, for several years and obtained accuracies greater than 90% for each consecutive year, using a hybrid classifier. These results were produced by the mean of nine images and highlighted the significance of the multi-temporal approach for crop classification. This study focuses on making use of large-set images having a temporal gap.

Crop monocultures are unusual in India, as they are in many other parts of the world. It indicates that the same crop is not cultivated close together; instead, this multi-crop method is commonly used. As a result, the specific-crop spectral-response number of times delivers the same spectral response as other crops. It may be due to their similar plantation date or due to the similarity in crop physiological-cycle. In such cases, single-date satellite imagery is generally not enough for mapping a specific crop of interest (Masialeti et al., 2010).

In the field of remote-sensing applications, the spectral vegetation-indices have been widely used for evaluating the biomass, water, plant, and crops (Jackson & Huete, 1991). Panigrahy et al. (2009) utilised the time-series AWiFS data to classify crops in accordance with their growing period. The Transformed divergence (TD) was computed as a separability measure, and amongst the various band combinations, the one having the maximum value of TD was selected as the best band combination to discriminate various classes of crops. It has been found that the band combination of Red, Near-Infrared (NIR) and Short-Wave Infrared (SWIR) bands depicted the maximum separability, and hence, after applying the Maximum Likelihood Classifier (MLC) scheme the overall accuracy got increased. This research concluded that selecting the proper band combination is essential while discriminating the various crop classes.

In some cases, multi-temporal data has very high dimensionality, which one way carries a large number of features. This problem has been referred to as the 'curse of dimensionality' in various studies. Here, the performance of the classifiers may be reduced due to an imbalance between the training samples and features(Liu & Sun, 2008). Temporal, multispectral data's spectral dimensionality can be reduced through deriving the vegetation index, to generate the temporal-indices database. The temporal-indices database in one way makes data suitable to be used with the existing classifiers to better process the feature vectors having only temporal dimension (Nandan et al., 2017).

Amongst the various vegetation indices generated so far, the NDVI (Rouse et al., 1974) and SAVI (Huete, 1988) have been widely used in multi-temporal studies. The NDVI studies biophysical characteristics by using reflectance values corresponding to the NIR and red wavelength portions of the electromagnetic spectrum. The red region absorbs chlorophyll, while in the near-infrared region, vegetation reflects quite well. The most typically and generally applied index for emphasising a vegetated area is the NDVI.

Bhandari et al., (2012) used advantages of NDVI feature extraction for the study area Jabalpur, India. They generated NDVI from the satellite images of the study area to find out the spectral signatures of various land covers such as vegetated area, water resources, concrete structures, road structures, and urban areas. It has been reported in their study that NDVI is very valuable for identifying ground features and it has wide applications in urban planning. This kind of vegetation analysis can provide useful information on natural disasters, which, in turn, can be utilised in developing protection strategies.

Gandhi et al., (2015) employed NDVI to identify vegetation categories such as scrub, agricultural, thick forest, thin forest-areas, and other places such as barren land and water bodies from multispectral imagery of the study area in Vellore district, Tamil Nadu, India. They employed the NDVI differencing technique to analyse the land-cover change of the research area for six years, from 2001 to 2006, and reported on the index's utility in predicting natural disasters.

Upadhyay et al., (2012) tested Class-Based Sensor Independent-Normalised Difference Vegetation Index (CBSI-NDVI), which is a sensor-independent indices approach for computing the spectral index. This CBSI-NDVI index can be defined using the function in Eq. (3.1):

$$\text{CBSI-NDVI} = min, max^{f\left\{\left(p_{1.......n}\right)_{r,c}\right\}} k, \tag{3.1}$$

where 'p' represents digital number (DN), 'c' and 'r' are column and row, respectively, for the location of class, 'n' represents number of features in the image, and 'k' represents class of interest.

For a given band combination, the CBSI-NDVI is computed as Eq. (3.2):

$$CBSI - NDVI = \frac{\left(Max - Min\right)}{\left(Max + Min\right)} \tag{3.2}$$

There is an operator which selects appropriate bands to be used for CBSI-NDVI indices, as a replacement of NIR and red bands. This is equivalent to the traditional indices case, so that the concerned class gets maximum enhancement. It depends upon the corresponding maximum and minimum DN-value bands for that particular class. Another advantage of adopting this method of index production is that the user is not required to select the bands to be utilised for a specific index (Upadhyay et al., 2012). With the large number of existing bands in a multi-spectral image, there are chances to have the best band-selection for any particular index while using a CBSI approach, as it consider all bands of a remote sensing image for a particular class of interest and hence selects the required combination of maximum and minimum DN-producing bands. Thus, CBSI reduces the spectral dimensionality without disturbing the temporal dimensionality. It is sensor-agnostic and based on a certain class, as the name implies. This strategy can be applied to any sensor data, since the algorithm selects the appropriate bands to utilise in the indices formula by employing a minimum and maximum bands selection approach. As a result, human intervention in band selection is not required. The requirement for finding CBSI-NDVI index is the location of class of interest in terms of latitude, longitude, or row and column values. At a selected class location, the algorithm picks up pixel values from all the present bands in an image, while minimum and maximum digital-number value-bands will be identified using minimum and maximum operators, which will give maximum spectral enhancement of that class. These minimum and maximum bands are used to determine the CBSI-NDVI index. A condition is also applied here such that: when the index value is negative, it will be replaced by zero. This strategy eliminates the need-to-know spectral information, such as which band is which, while searching for a relevant index database. CBSI essentially substitutes traditional bands with maximum and minimum band-values for a class.

One of the primary criteria in time-series studies is to choose the best date combination for distinguishing between class of interest and non-target classes. Thus, choosing the appropriate temporal date-combination from available temporal data is critical in these investigations; otherwise, categorisation will become rigid, reducing accuracy. Murthy et al. (2003) employed the Bhattacharya distance metric to determine the separability of wheat and other crops using three datasets with only one, two, or three overpass dates, respectively. They concluded that the three-date combination showed the highest spectral separability, as well achieving the highest classification accuracy amongst all the three datasets.

According to Hajj et al. (2007), the condition for time-series remote-sensing data-acquisition should be rain, cloud, and haze-free. In a number of instances, proper optical time-series data is not available due to unfavourable atmospheric circumstances, as well as the satellite's coarser repeat-cycle, and so on. As a result, there are always gaps in temporal data (Steven et al., 2003). Therefore, a single sensor may not be helpful enough in temporal studies. These problems are easily overcome by combining other optical remote-sensing data or by using microwave data to fill in the temporal gaps. In the current context, numerous optical and SAR sensor-based data from various satellites are available. The optical remote-sensing satellites can be LandSat-8/9, Sentinel-2A/2B, PlanetScope, and so on, while the SAR remote sensing images can be from Sentinel-1A/1B, Radarsat-1, IRS-RISAT and so on. The well-known benefits of SAR images are owing to their applicability in both cloud and night settings. The backscatter coefficient or SAR-based index or any other parameter from SAR remote sensing images might be considered when integrating SAR images with optical views. Various indices can be utilised while integrating optical with optical satellite-temporal data. There may arise the issue of different pixel size, which can be resolved while bringing all the data pixels at the same spatial level.

Vincent et al. (2020) experimented with a modified temporal-indices database generated from SR, Chlorophyll Index (CI), SAVI, and NDVI using the weighted red-edge. Here, in this study, the reflectance of the red-edge region has been added with the red-band reflectance in a controlled way and replaced in the computation of vegetation indices. Like in NDVI, the red-edge band has been added with different weight constants such as 0.1, 0.4, and 0.7 as the red-edge weight constant, or controlling parameter (a) value. Later this parameter 'a' is extended to 0.5, 0.6, and 0.8 and then NDVI was further computed. Finally, SR, CI, and SAVI computations have been done with an 'a' weight value of 0.7. Each derived vegetation indices then stacked them together for all the dates considered. Hence, temporal vegetation indices were generated corresponding to the phenological cycle of the crop. It has been discovered that considering the crop's phenological information is critical for single-crop identification. The stage of the plant in its phenological cycle is determined by the background soil. As a result of the appearance of background soil in time domain with plant stage, it is possible to distinguish between various vegetation species.

Chhapariya et al. (2021a, 2021b, 2020) worked on mapping burnt paddy fields while applying kernel function to handle the non-linearity. In stubble paddy burning, two dates' temporal data has been used to map the burnt paddy fields in the surroundings of the district Patiala, Punjab, India on 24 October 2019 and 25 October 2019. The temporal data used in this study was from Sentinel-2, which is capable of providing the images in an interval of around five days. From this case study, it was concluded that the burnt paddy fields can be mapped at an interval of five to ten days. While two dates temporal data was used to avoid spectral overlap between burnt paddy fields and other classes like roads, open ground without vegetation, settlements etc.

3.3 SPECIFIC SINGLE CLASS MAPPING

There are large number applications where the single-class information is required. For example, these can be agriculture areas such as specific crop fields or acreage, specific forest species and fields affected due to some calamity. All such types of class mapping come under level-two or level-three classification. There is a problem of spectral overlap when undertaking mapping at these levels; for example, identifying wheat within agriculture fields or identifying infected crops or damaged crop fields. This challenge is easily handled with temporal data, since it can provide a unique time-domain-based signature of a certain class of interest. On the other hand, using temporal, multispectral data, two dimensions-spectral and temporal-need to be handled, as most of the classification models are capable of taking only one of them (Nandan et al., 2017). To sort out these issues, various pre-processing of digital-image processing techniques must be used to minimise the spectral dimension while retaining the temporal dimension. To minimise the spectral dimensions of temporal, multi-spectral remote-sensing data, the indices approach works best. By band ratioing, the indices approach enhance the class of interest. As a result, using this approach during the pre-processing stage can result in a database of temporal indices. This database can provide one-of-a-kind temporal-domain information regarding a particular class of interest.

3.4 INDICES FOR TEMPORAL DATA

As discussed earlier, temporal indices play very important role while processing temporal, optical, thermal and SAR remote-sensing data. Here, enhancing a class by an index is the main application. However, it also reduces the spectral dimensionality of optical remote-sensing data. The vegetation information in remotely sensed images is mostly interpreted because of the distinct spectral behaviour of plant green leaves or canopy from other land covers in the red and near-infrared regions of the electromagnetic spectrum. Further, working with the red-edge region and generating improved indices has produced good results strongly related to the physiological status of plants. The NDVI is a widely used vegetation index to estimate biophysical variables, and it relies on chlorophyll absorption in the red band and high reflectance in the NIR band, while the range of NDVI goes from −1 to +1.

Agriculture methods are not monoculture in many countries across the world. In addition to crop techniques, the planting date and phenological stage provide spectral similarity between classes (Misra et al., 2012). Since the time-series data is capable of incorporating the changes occurring during the different stages of crop growth. Therefore, use of temporal remote-sensing data is capable of identifying the specific crop or vegetation class, as it is capable of handling the overlap of the spectral response of specific crops with other classes or other vegetation classes. Masialeti et al. (2010) analysed the temporal NDVI curves for 2001, as well as 2005 datasets. The goal of this study was to determine the NDVI profile from MODIS data for the primary crops in the Kansas region that had a consistent crop cycle, which was assessed graphically and statistically. It was also investigated, to use the possibility of 2001 dataset to map the 2005 crop.

Wardlow et al. (2007) considered a 12-month, temporal MODIS NDVI and EVI database to study the agricultural crop fields. It has been identified that most of the crops can easily be separated from each other during their growing season and also the interclass variability of the crops has influenced the spectral separability. It may be due to different planting dates and climate. Further, a comparison of the performance of these two indices showed that there has been a strong correlation between them during the growing season of the crop.

To generate the NDVI from multispectral image selection of appropriate bands is required. Li et al. (2010) clearly indicated the importance of the proper selection of bands. In the study, to discriminate amongst crops, four temporal MODIS NDVI images were generated. It has been concluded in this study that the proper selection of bands improves classification accuracy. Another research study by Simonneaux and Francois (2003) used temporal vegetation indices for the classification of SPOT images. They studied the impact of temporal images on classification accuracy. It can be concluded from this study that there is an impact on the accuracy of classification with the varying number of temporal images. In the study, the accuracy of classification has been increased by 2% from the earlier one, when five temporal images were used to cover the whole phenology.

The spectral reflectance curve for vegetation represents a peak (strong reflection) and valley (predominant absorption) configuration. The chlorophyll pigment within the chloroplasts (and other pigments) in palisade cells absorbs blue and red wavelengths, as it is required for photosynthesis and reflects back 10–15% of green light. Infrared radiation, on the other hand, penetrates the leaf as far as mesophyll cells, where it is reflected back by nearly half—a phenomenon known as internal scattering at the cell wall-air interface, with the remaining energy transmitted or reflected by leaves beneath as leaf-additive reflectance. Plant stress is regularly responsible for enhanced reflectance in the visible zone. Only when stress created is strong enough to cause severe leaf dryness does IR reflectance drop (Jensen, 1996).

Reflectance at red edge provides the unique response of the green vegetation and can be used to distinguish them from the other targets. This particular response is known as red-edge property, and it results from internal leaf-scattering (large NIR reflectance), as well as chlorophyll absorption (low red reflectance)—two essential optical characteristics of plant tissue (Boochs et al., 1990).The red-edge bands can be effectively utilised in detecting small changes in morphological and chemical properties of plants (Boochs et al., 1990). The relation between the shape of the red-edge band and the wavelength for wheat crops has been studied. It has been identified that the inflection point is due to variation in the chlorophyll content of the plant. There has been a shift of the red-edge band towards longer wavelengths, when chlorophyll content increases. Overall, spectral slope in the red-edge portion of the reflectance curve is dependent on chlorophyll concentration, as well as on additional effects, such as growth stage, species, leaf stacking, and leaf water-content. The application of red-edge spectral information is important for knowing chlorophyll significance, as well as leaf-area index estimation. Red-edge information is not much dependent on ground-cover variation but very much suitable for detection of early crop-stress. It has been noted that, as chlorophyll levels in vegetation rise, so does the red edge, which shifts to longer wavelengths. As a result, the red-edge position can be used to calculate the chlorophyll concentration of a leaf. When chlorophyll content declines, the red edge shifts to shorter wavelengths. If there is leaf stacking, the concentration of chlorophyll increases. Red-edge band information shifts to longer wavelengths as a result of structural changes in leaf tissues or leaf optical-characteristics (Horler et al., 1983).

The radiation propagation through leaves can be modelled using Beer Lambert's law, which includes absorption as well as a scattering coefficient. While following this law, it has been noticed that, with respect to the concentration, the transmittance spectrum shifts. Path length increases as a result of scattering, resulting in higher absorbance. At a given concentration, overall internal leaf structure, non-homogeneous pigment concentration, and scattering affect leaf reflectance. Light is sufficiently backscattered when passing through a significant number of layers of leaves. As the layer size increases beyond 725nm, the reflectance increases. As the number of leaves in this stack increases, the reflectance shifts to longer wavelengths. This is due to dispersion, which raises the effective pigment-concentration by increasing the path length. As a result, the absorption and scattering coefficients can distinguish between growth stages, plant species, and leaf water-content (Horler et al., 1983).

Horler (1983) studied maize leaves and reported that red-edge measurements can be effectively utilised in predicting the status of chlorophyll and early stress-assessment of plants. They explained the relationship of wavelength shift and chlorophyll content using the Beer-Lambert law.

Multispectral band-ratioing decreases the spatial and temporal changes in reflected light induced by geometric factors such as topography. The differences in spectral properties of green leaves cause the interpretation of vegetation in remote-sensing images. The vegetation index is a radiometric measure that serves as an indicator of the relative abundance and activity of green vegetation. When analysing high spectral-reflectance in the red-edge region using waveform-data mode, variability is decreased, which is associated with broad-band data. When the red band in the indices is joined with the red-edge band, it closely correlates with the physiological status of the plant.

3.5 TEMPORAL DATA WITH MULTI-SENSOR CONCEPT

It is now clear that temporal studies are very helpful for studying the unique behaviour of plants and are capable of extracting specific classes. However, generating the temporal database using a cloud-free, optical remote-sensing image to cover the whole phenology of the crop is a matter of concern. The temporal or periodical monitoring of various crops during their different growth stages help it to discriminate them from others. The unique spectral signature of a crop during its growth stage is known as its phenology, which helps it to identify amongst the various other vegetation species or crop types (El Hajj et al., 2007). Due to the sensitivity of optical remote-sensing to atmospheric conditions, temporal gaps in crop studies may emerge due to the non-availability of optical remote-sensing images from a specific sensor. The cloud-free optical data is essential to consider crop phenology to study specific crop or specific vegetation species. In a temporal study conducted by Steven, the accuracy of the classification has been decreased with a decrease in the amount of sufficient temporal-date data. A study conducted by Mcnairn et al. (2005) using Landsat and SPOT datasets integrating with RADARSAT-1 as well as Envisat-ASAR for analysing crop inventory on annual basis has applied a maximum-likelihood classifier and achieved accuracy up to 80%. Shang et al. (2008) conducted another research study using multi-sensor and multi-temporal data using a decision-tree classification. The accuracy obtained by this classification was 87%, indicating that accuracy increases as they integrate microwave data with optical remote-sensing data.

The temporal vegetation indices are generated by applying mathematical operations using two or more bands and then stacking in the temporal domain. This reduces the spectral dimensionality, as well as enhancing the vegetation signal while normalising sun angle. According to Aggarwal (2015), it is found that the temporal-indices database has provided unique crop-phenological information to discriminate and map single crops, and the accuracy of single-crop mapping results depends upon the usage of optimum, temporal remote-sensing data dates. Van Niel and McVicar (2004) conducted a similar study to identify the best temporal window with an increase in overall accuracy. The maximum-likelihood classifier was utilised in the study to establish the appropriate temporal window for single-class identification with better crop separating and increased overall accuracy.

Zurita-Milla et al. (2011) tried to study the usefulness of single and multi-temporal images while applying spectral un-mixing methods on MERIS data for land-cover mapping. From this study, it was found that multi-temporal data was better to discriminate classes than single dates when in an area where classes are mixed.

Navarro et al. (2016) experimented with SPOT-5 and Sentinal-1A data for crop-monitoring information-generation and for crop water-requirement estimation. Sentinel-1A images were integrated with SPOT-5 images to refine temporal resolution to use this data in agriculture applications. In the research work, NDVI from SPOT-5 data and dual backscatter-coefficients from Sentinel-1A images time-series data that were used to compute K_{cb} (Basal Crop Coefficient) curve. K_{cb} curve was generated for four different types of crops. From the study, it has been observed that NDVI and the backscatter coefficient provide unique information as time-series data for discriminating major types of crops present in the study area. Limitations of the research work were attributed to insufficient temporal EO data that every stage of the crop cycle was not able to identify.

As it is now clear that temporal optical/microwave data-sets from different sensors, including multi-spectral, hyperspectral, thermal remote-sensing, and SAR, can be integrated to generate indices/backscatter coefficients from these data sets.

Moreover, if their spatial resolution parameters are different, then the coarser resolution data-sets can be resampled to the finest spatial-resolution data-sets to make a pixel-to-pixel correspondence. If the radiometric resolutions of various optical sensors disagree, this will have no effect because the radiometric discrepancy will be for the entire image of that sensor and date. Further, the different spectral resolution of different sensor data may also not affect generating a temporal-indices database. While integrating microwave data with optical data, the indices/backscatter-coefficient curve (Figure 3.1) between point a and b may go up or down and may not follow as per the single, sensor-based indices curve. However, it still represents the variation of stages of a crop or vegetation during a and b points.

There may also arise a question about how the backscatter coefficient or indices from SAR images can be integrated with indices from optical images. To integrate SAR images with optical images, the indices generated from the optical can be brought to an identical scale as the backscatter coefficient or SAR-based indices using equation 3.3. Also, the pixel size of all the sensors should be brought equal to the finer pixel size.

$$\bar{X} = ai + bj + ck + dl \tag{3.3}$$

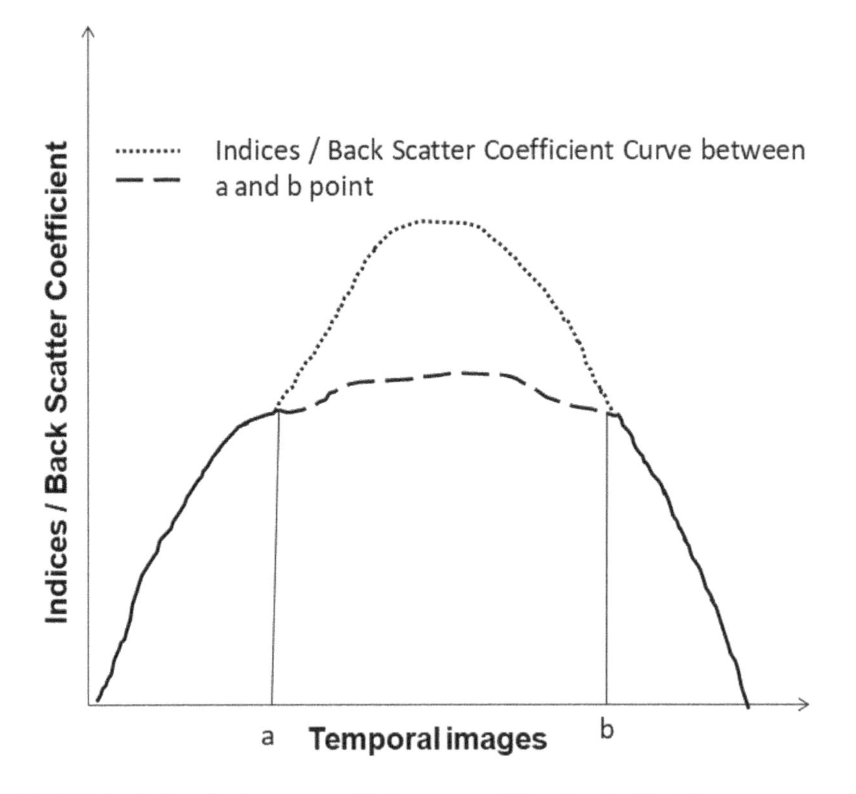

FIGURE 3.1 Indices/backscatter-coefficient curve while using dual/multi-sensor, temporal data-set.

where a, b, c can be indices from optical images and d can be the backscatter coefficient or indices from SAR temporal images.

There are several websites where one can download multi-sensor remote-sensing data-sets, such as https://scihub.copernicus.eu/ for Sentinel-1A/1B and Sentinel-2A/2B data sets. To download the Landsat series data-set the website available is https://earthexplorer.usgs.gov/. Planetscope/Dove satellite data at 3m with daily temporal frequency is available at www.planet.com/products/. To explore RADARSAT data sets one can go to https://earth.esa.int/eogateway/missions/radarsat. MAXAR high-resolution data is available from the website www.maxar.com/. One may explore www.cloudeo.group/ for extremely high-resolution remote-sensing images. The website https://skywatch.com/ brings all remote-sensing image-providers together on one website.

3.6 SUMMARY

This chapter has discussed about the methods to create a temporal database, while keeping the importance of the indices approach in mind. An information about integrating dual and multi-sensor temporal datasets to make them useful for specific-class extraction have also been discussed. The next Chapter 4 discusses about training techniques, training sample size, and its importance in fuzzy and deep learning classifiers.

BIBLIOGRAPHY

Aggarwal, R. (2015). *Specific Crop Identitifcation Using Kernel*. Indian Institute of Remote Sensing.

Bhandari, A. K., Kumar, A., & Singh, G. K. (2012). Feature extraction using Normalized Difference Vegetation Index (NDVI): A case study of Jabalpur City. *Procedia Technology*, *6*, 612–621. https://doi.org/10.1016/j.protcy.2012.10.074.

Boochs, F., Kupfer, G., Dockter, K., & Kuhbaüch, W. (1990). Shape of the red edge as vitality indicator for plants. *International Journal of Remote Sensing*, *11*(10), 1741–1753. https://doi.org/10.1080/01431169008955127.

Chhapariya, K., Kumar, A., & Upadhyay, P. (2021a). Kernel-based MPCM algorithm with spatial constraints and local contextual information for mapping paddy burnt fields. *Journal of the Indian Society of Remote Sensing*, *49*. https://doi.org/10.1007/s12524-021-01346-1.

Chhapariya, K., Kumar, A., & Upadhyay, P. (2020). Handling non-linearity between classes using spectral and spatial information with kernel based modifiedpossibilistic*c*-means classifier. *Geocarto International*, 37:6, 1704-1721, https://doi.org/10.1080/10106049.2020.1797186.

Chhapariya, K., Kumar, A., & Upadhyay, P. (2021b). A fuzzy machine learning approach for identification of paddy stubble burnt fields. *Spatial Information Research*, *29*, 319–329. https://doi.org/10.1007/s41324-020-00356-4.

El Hajj, M., Bégué, A., & Guillaume, S. (2007). Multi-source information fusion: Monitoring sugarcane harvest using multi-temporal images, crop growth modelling, and expert knowledge. *Proceedings of MultiTemp 2007–2007 International Workshop on the Analysis of Multi-Temporal Remote Sensing Images*. https://doi.org/10.1109/MULTITEMP.2007.4293064.

Gandhi, G. M., Parthiban, S., Thummalu, N., & Christy, A. (2015). Ndvi: Vegetation change detection using remote sensing and gis—A case study of vellore district. *Procedia Computer Science*, *57*, 1199–1210. https://doi.org/10.1016/j.procs.2015.07.415.

Jackson, R., & Huete, A. (1991). Interpreting vegetation indices. *Preventive Veterinary Medicine*, *11*, 185–200. https://doi.org/10.1016/S0167-5877(05)80004-2.

Jensen, J.R. (1996) *Introductory Digital Image Processing: A Remote Sensing Perspective*. 2nd Edition, Prentice Hall, Inc., Upper Saddle River, NJ.

M. el Hajj, A. Begue and S. Guillaume, "Multi-source Information Fusion: Monitoring Sugarcane Harvest Using Multi-temporal Images, Crop Growth Modelling, and Expert Knowledge," 2007 *International Workshop on the Analysis of Multi-temporal Remote Sensing Images*, Leuven, Belgium, 2007, pp. 1-6, doi: 10.1109/MULTITEMP.2007.4293064.

Horler, H., Dockray, M., & Barber, P. J. (1983). The red edge of plant leaf reflectance. *International Journal of Remote Sensing*, *4*(2), 273–288. https://doi.org/10.1080/01431168308948546.

Huete, A. (1988). A soil-adjusted vegetation index (SAVI). *Remote Sensing of Environment*, *25*(3), 295–309. https://doi.org/10.1016/0034-4257(88)90106-X

Li, Y., Chen, X., Duan, H., & Meng, L. (2010). An improved multi-temporal masking classification method for winter wheat identification. In *ICALIP 2010–2010 International Conference on Audio, Language and Image Processing, Proceedings*, Shanghai, China, 2010, pp. 1648–1651. https://doi.org/10.1109/ICALIP.2010.5685073

Liu, Z., & Sun, Z. (2008). Active one-class classification of remote sensing image. *International Conference on Earth Observation Data Processing and Analysis (ICEODPA)*, *7285*(January), 72850L. https://doi.org/10.1117/12.816115

Kriegler, F., Malila, W., Nalepka, R., & Richardson, W. (1969). Preprocessing transformations and their effect on multispectral recognition. In *Proceedings of the 6th International Symposium on Remote Sensing of Environment*. Ann Arbor, MI: University of Michigan, pp. 97–131.

Masialeti, I., Egbert, S., & Wardlow, B. D. (2010). A comparative analysis of phenological curves for major crops in Kansas. *GIScience & Remote Sensing*, *47*(2), 241–259. https://doi.org/10.2747/1548-1603.47.2.241

McNairn, H., Shang, J., Champagne, C., Huffman, E., Smith, A., & Fisette, T. (2005). A multi-sensor approach to inventorying agricultural land use. In *Proceedings of the 31st INternational Symposium on Remote Sensing of Enviroment*, pp. 0–3. Retrieved from www.isprs.org/proceedings/2005/ISRSE/html/papers/743.pdf

Misra, G., Kumar, A., Patel, N. R., & Singh, A. (2012). Mapping specific crop-A multi sensor temporal approach. *IGARSS*, 3034–3037.

Murthy, C. S., Raju, P. V., & Badrinath, K. V. S. (2003). Classification of wheat crop with multi-temporal images: Performance of maximum likelihood and artificial neural networks. *International Journal of Remote Sensing*, *24*(23), 4871–4890. https://doi.org/10.1080/0143116031000070490.

Nandan, R., Kumar, R., Kumar, A., & Kumar, S. (2017). Wheat monitoring by using kernel based possibilistic *c*-means classifier: Bi-sensor temporal multi-spectral data. *Journal of the Indian Society of Remote Sensing*, *45*(6), 1005–1014. https://doi.org/10.1007/s12524-016-0651-9.

Navarro, A., Rolim, J., Miguel, I., Catalão, J., Silva, J., Painho, M., & Vekerdy, Z. (2016). Crop monitoring based on SPOT-5 Take-5 and sentinel-1A data for the estimation of crop water requirements. *Remote Sensing*, *8*(6). https://doi.org/10.3390/rs8060525.

Panigrahy, R. K., Ray, S. S., & Panigrahy, S. (2009). Study on the utility of irs-p6 AWiFS SWIR Band for crop discrimination and classification. *Journal of the Indian Society of Remote Sensing*, *37*(2), 325–333. https://doi.org/10.1007/s12524-009-0026-6.

Sakamoto, T., Yokozawa, M., Toritani, H., Shibayama, M., Ishitsuka, N., & Ohno, H. (2005). A crop phenology detection method using time-series MODIS data. *Remote Sensing of Environment*, *96*(3–4), 366–374. https://doi.org/10.1016/j.rse.2005.03.008.

Sengar, S. S., Kumar, A., Ghosh, S. K., & Wason, H. R. (2012). Soft computing approach for Liquefaction identification using landsat-7 temporal indices data. *International Archives of the Photogrammetry, Remote Sensing and Spatial Information Sciences.* XXXIX-B8, 61–64, https://doi.org/10.5194/isprsarchives-XXXIX-B8-61-2012.

Serra, P., Pons, X., & Saurí, D. (2008). Land-cover and land-use change in a Mediterranean landscape: A spatial analysis of driving forces integrating biophysical and human factors. *Applied Geography, 28*(3), 189–209. https://doi.org/10.1016/j.apgeog.2008.02.001

Shang, J., McNairn, H., Champagne, C., & Jiao, X. (2008, July). Contribution of multi-frequency, multi-sensor, and multi-temporal radar data to operational annual crop mapping. In *Geoscience and Remote Sensing Symposium, 2008. IGARSS 2008. IEEE International* Boston, MA, USA,(Vol. 3, pp. III–378). IEEE.

Simonneaux, V., & Francois, P. (2003). Identifying main crop classes in an irrigated area using high resolution image time series. *Geoscience and Remote Sensing Symposium, 2003. IGARSS '03. Proceedings. 2003 IEEE International, 1*(October), 252–254. https://doi.org/10.1109/IGARSS.2003.1293741.

Steven, M. D., Malthus, T. J., Baret, F., Xu, H., & Chopping, M. J. (2003). Intercalibration of vegetation indices from different sensor systems. *Remote Sensing of Environment, 88*(4), 412–422. https://doi.org/10.1016/j.rse.2003.08.010.

Tucker, C. J. (1979). Red and photographic infrared linear combinations for monitoring vegetation. *Remote Sensing of Environment, 8*(2), 127–150. https://doi.org/10.1016/0034-4257(79)90013-0.

Upadhyay, P., Kumar, A., Roy, P. S., Ghosh, S. K., & Gilbert, I. (2012). Effect on specific crop mapping using WorldView-2 multispectral add-on bands: Soft classification approach. *Journal of Applied Remote Sensing, 6*(1), 063524-1–063524-15. https://doi.org/10.1117/1.jrs.6.063524.

Van Niel, T. G., & McVicar, T. R. (2004). Determining temporal windows for crop discrimination with remote sensing: A case study in south-eastern Australia. *Computers and Electronics in Agriculture, 45*(1–3), 91–108. https://doi.org/10.1016/j.compag.2004.06.003.

Vincent, A., Kumar, A., & Upadhyay, P. (2020). Effect of red-edge region in fuzzy classification: A case study of sunflower crop. *Journal of the Indian Society of Remote Sensing, 48.* https://doi.org/10.1007/s12524-020-01109-4.

Wardlow, B. D., Egbert, S. L., & Kastens, J. H. (2007). Analysis of time-series MODIS 250m vegetation index data for crop classification in the U.S. Central Great Plains. *Remote Sensing of Environment, 108*(3), 290–310. https://doi.org/10.1016/j.rse.2006.11.021.

Wardlow, B. D., & Egbert, S. L. (2008). Large-area crop mapping using time-series MODIS 250 m NDVI data: An assessment for the U.S. Central Great Plains. *Remote Sensing of Environment, 112*(3), 1096–1116. https://doi.org/10.1016/j.rse.2007.07.019.

Zurita-Milla, R., Gómez-Chova, L., Guanter, L., Clevers, J. G., & Camps-Valls, G. (2011). Multitemporal unmixing of medium-spatial-resolution satellite images: A case study using MERIS images for land-cover mapping. *IEEE Transactions on Geoscience and Remote Sensing, 49*(11), 4308–4317. https://doi.org/10.1109/TGRS.2011.2158320.

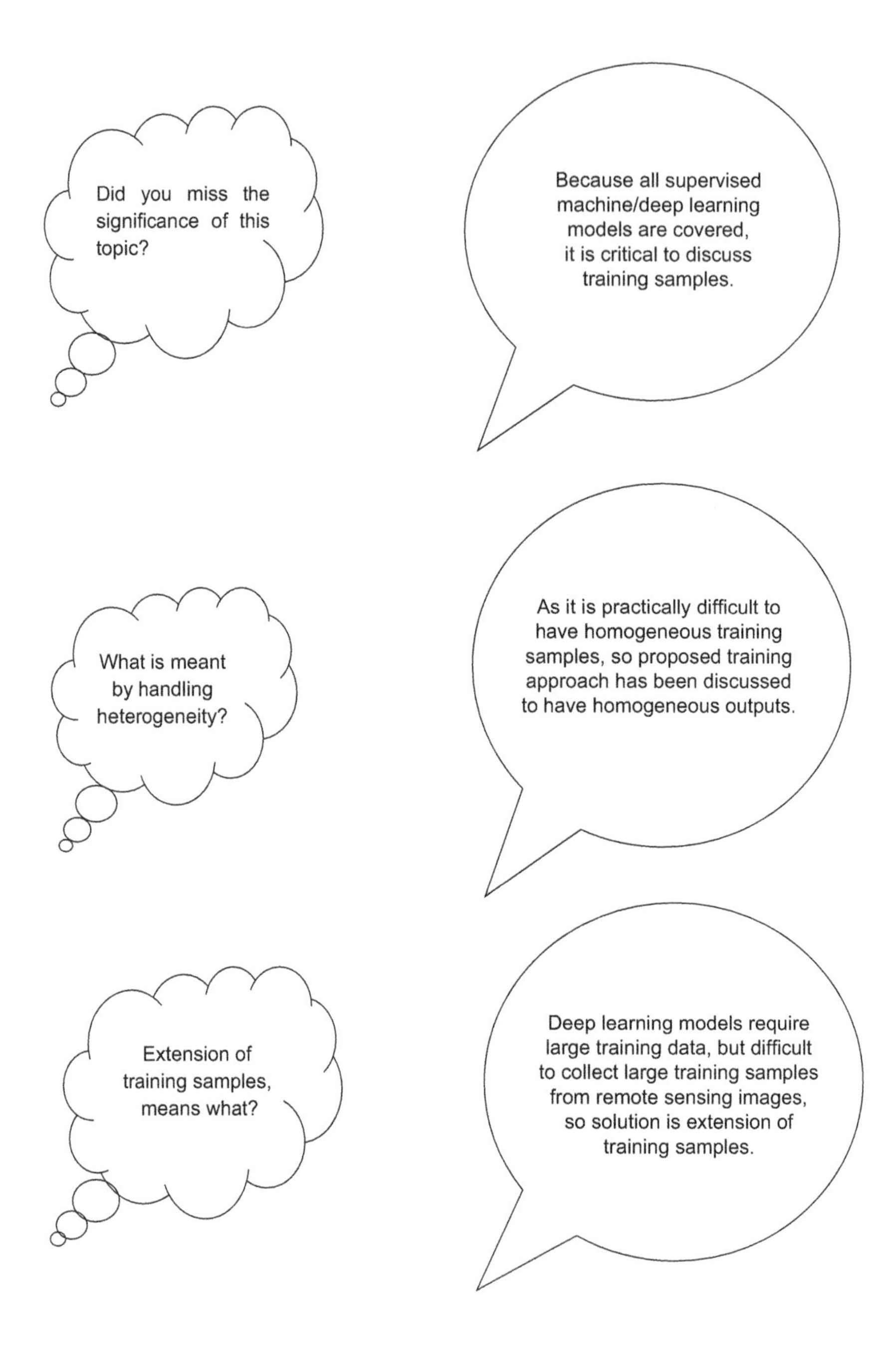

No Gain. . . .
Without Pain. . . .

4 Training Approaches—Role of Training Data

4.1 INTRODUCTION

The role of training data is to generate the classifier parameters or to tune the learning weights (Aaron et al., 2018). However, before using training data, it is necessary to understand the conditions that must be taken into account when collecting training data. Training samples should be evenly distributed or be homogeneous, and they should accurately reflect a class. However, due to heterogeneity within class, training samples cannot be homogeneous. Nowadays, there is a large number of machine-learning classifiers and their requirement of samples size is of a different nature (Yuji et al., 2018). For statistical varying classifiers, the statistical parameters are generated from training samples. However, the heterogeneity within the class cannot be handled while using statistical samples generated from the training data. There are different criteria for deciding the training-data size. The overall application of training samples is to generate required parameter-values for statistical classifiers, while for learning-classification algorithms, weights and other parameters need to be tuned.

Ground-truth data-sampling comes under the statistical domain for selecting, manipulating, and analysing a demonstrative subset of data-sample points. It extracts patterns and provides a base for learning larger database. Sample data-sets provide academicians, scientists, analytical modellers, and others the means, to analyse the small and manageable amount of data representing the population. Thus, it is used to develop and analyse models more easily on existing machines in a short time and produce meaningful findings (Eric et al., 2021).

There are various methods for collecting samples from a given data or from the ground. The best method depends upon the data set and requirement. Data samples can be collected based on probability, which uses random numbers. More variants in probability sampling include (Hamed, 2016):

a) *Simple random sampling:* Through random function, the random samples are selected from the whole population.
b) *Stratified sampling:* In this approach, subsets of the data sets are created based on a common criterion, and then random samples are collected from each subset.
c) *Cluster sampling:* In this method, a larger data set is divided into small sets based on a defined criterion; later, random sampling is done from clusters.
d) *Multi-stage sampling:* This is a more complicated form of cluster sampling. Here, the second-level clustering is done based on some criteria. The second-level clusters were analysed to collect samples.
e) *Systematic sampling:* Through this approach, samples are picked at an interval from a large population of databases.

DOI: 10.1201/9781003373216-4

f) *Non-probability sampling:* In this approach, samples are selected based on analyst judgement on given population data. Since an analyst selects samples, it is, therefore, difficult to extrapolate about all samples accurately to represent a larger population, in comparison to the probability-sampling method.

4.2 HANDLING HETEROGENEITY WITHIN A CLASS

One of the categories of machine-learning algorithms generates statistical parameters from a given-samples data-set. The mean statistical parameter is one of the common parameters generated from training-sample data. This parameter does not represent a full range of sample data; however, it is actually close to anyone's sample value. As a matter of fact, as the mean parameter used in the machine-learning algorithm, a lot of sample points belonging to the same class remains far from the parameter criteria considered for classification. Therefore, these data points do not get a label for that class. Thus, the mean parameter-factor is not able to handle the heterogeneity within a class. Heterogeneity actually represents when there is variability in the dataset or within a class, which means that sample data are different. It is the opposite of homogeneity (Giacomo et al., 2012). Homogeneity and heterogeneity are perception statistics correlating to uniformity in a given data set.

Break time to think!

While collecting training samples. . . .
. . . .Three points to consider:

Samples should be well distributed—this can be ensured.
Samples should truly represent the class—care should be taken.
Samples should be homogeneous—cannot be taken care of in the case of remote-sensing images.
(Water is the most homogeneous class, while urban is the most heterogeneous class.)
So, heterogeneity in training samples depends on samples collected for which class?
Solution—think over using the ISM (Individual Sample as Mean) training approach.

A homogeneous data is uniform in configuration while heterogeneous data is clearly non-uniform. For example, in a remote-sensing image, digital values within wheat fields have a large range of values.

Using this concept, the wheat fields are heterogeneous with respect to digital values. It is due to the heterogeneity present in a given data set, and hence, cannot be handled while using the mean parameter generated from training data. In place of generating the mean parameter from sample data, each sample value can be treated as mean, to be used in a statistical classifier as shown in Figure 4.1, called the 'individual sample as mean' (ISM) training-parameter approach. This approach can handle heterogeneity within a class while considering the impact of each sample on classification outputs.

Heterogeneity within class – considering each sample impact

Not seen in machine learning algorithms as they are implemented

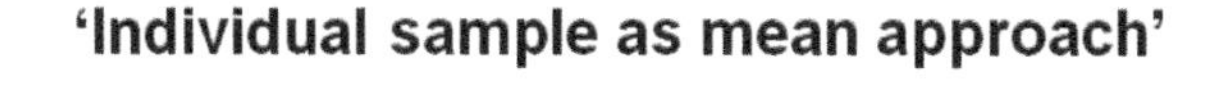

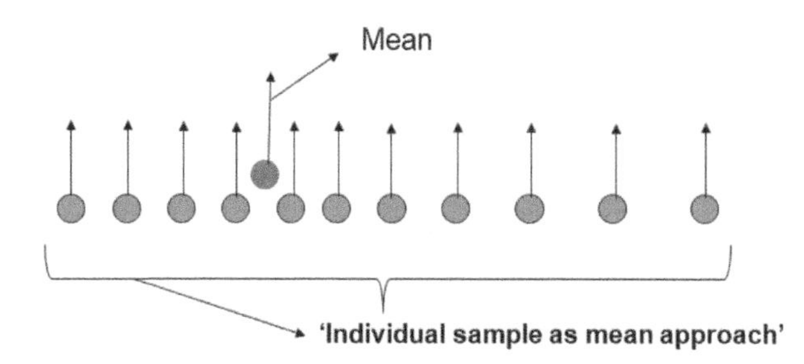

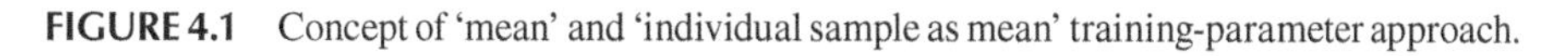
FIGURE 4.1 Concept of 'mean' and 'individual sample as mean' training-parameter approach.

Statistic represents data distribution and patterns….
….But true representation of class through samples is lost?

4.3 MANUAL OR REGION GROWING METHOD FOR TRAINING-SAMPLES COLLECTION

Training samples from remote-sensing images can be collected manually or by an automatic approach using some similarity or dissimilarity algorithms. In the manual method, training pixels from remote-sensing images can be collected in point, line, or polygon mode. While collecting training pixels manually, there is a chance that samples are not homogeneous and, a number of times due to human error, samples may not represent the true class label. To avoid human error so that samples are truly representing class as well as maintaining homogeneity, training samples may be collected through the region growing method (Mubarak et al., 2012; Chaturvedi et al., 2016). In the region growing method, a seed-sample pixel is provided. Using the seed pixel within a defined search window, similar training pixels can be collected using similarity or dissimilarity algorithms-based criteria while defining the acceptable correlation threshold. The advantage of the region growing method is that training pixels truly represent the class. Also, heterogeneity within training pixels is reduced.

4.4 EXTENSION OF TRAINING SAMPLES

A number of times, due to various factors, it's not possible to collect a large size of training samples for a class. These factors can be due to less fund availability, the study site being in remote areas, or due to pandemic situation like covid, etc. There are machine/deep-learning models, which requires training data of a large

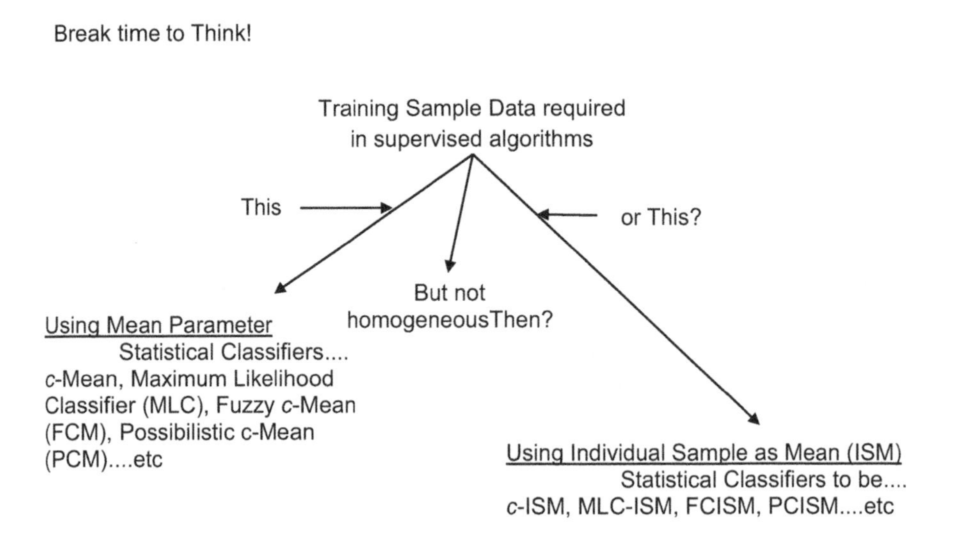

M or ISM....
Think over it?

To have homogenous classified output
Think over it?

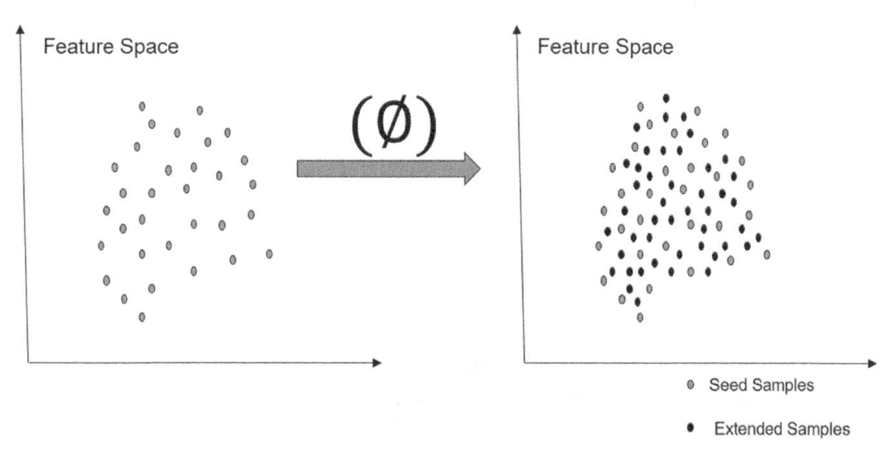

FIGURE 4.2 Extension of training samples.

size. When training sample size is smaller but machine/deep-learning models require large training-data, in that case, a proposed extension of the training sample approach can be applied. In an extension of the training sample approach using sample boundary information, new samples can be created to increase training-sample size (Figure 4.2). The advantage of increasing training sample size is that machine/ deep learning models can achieve a defined, acceptable training error threshold. Secondly, homogeneity is also achieved in classified output while applying training samples as the 'individual sample as mean' approach.

4.5 COGNITIVE APPROACH TO TRAIN CLASSIFIER

Cognitive science is the study of the ability of human brain to process the information, while taking previous experiences into account (Verwey, 2015). Cognitive science aids in the comprehension and processing of information. According to one definition, cognitive science is a subset of Artificial Intelligence. Cognitive science combines, for example, human sentiments and thoughts in machines for applications such as pattern recognition and data processing. Cognitive science, in essence, mimics human memory, learning mode, perception-based understanding, human IQ, and so on (Abdi, 2016). Cognitive science encompasses modern techniques for incorporating human experience as knowledge.

It is novel to apply cognitive science as cognitive image-processing in the domain of image-processing applications. A general trend of transitioning from computational-data processing to cognitive-information processing has been discovered, affecting many scientific areas (Diamant, 2014). The goal of this section is to the question, how to use cognitive science in temporal, remote-sensing image-processing to generate the training data. The generated training data should be useful for processing temporal remote-sensing data in the past or future using a supervised classification approach. The benefit of using cognitive science to generate training data for processing temporal remote-sensing data at some time interval is that it reduces the time and cost of going to fields on a regular basis to collect training samples.

Multi-temporal remote-sensing images are widely used to monitor specific classes of events that occur at regular intervals. Examples of monitoring specific classes occurring on the ground include harvested fields of a crop at regular intervals, stubble burnt paddy fields at regular intervals, and so on. As classifier models work in supervised mode, ground-truth training samples are required at each date to map specific class that occur at specific intervals. When the frequency of occurrence of a specific class to be mapped is high, training samples are required for each occurrence. Based on the current situation in terms of time and cost, collecting ground-truth training samples multiple times may not be feasible. Mragank et al. (2021) proposed a cognitive approach to training classifiers to overcome issues with the lack of frequent training data. Figure 4.3 shows an example of how a cognitive approach can be used to train a classifier model using seed-training samples in the forward or backward direction in the time domain.

From Figure 4.3, a seed-training sample has been collected during 26th October as stubble burnt fields. To avoid spectral overlap, the same, filed as dry stubble, has been considered from a 25th October image. Images of training samples of dry and burnt stubble were collected on 25th and 26th October. Now, dry-stubble training samples have been applied to the 26th October image and burnt-stubble training samples have

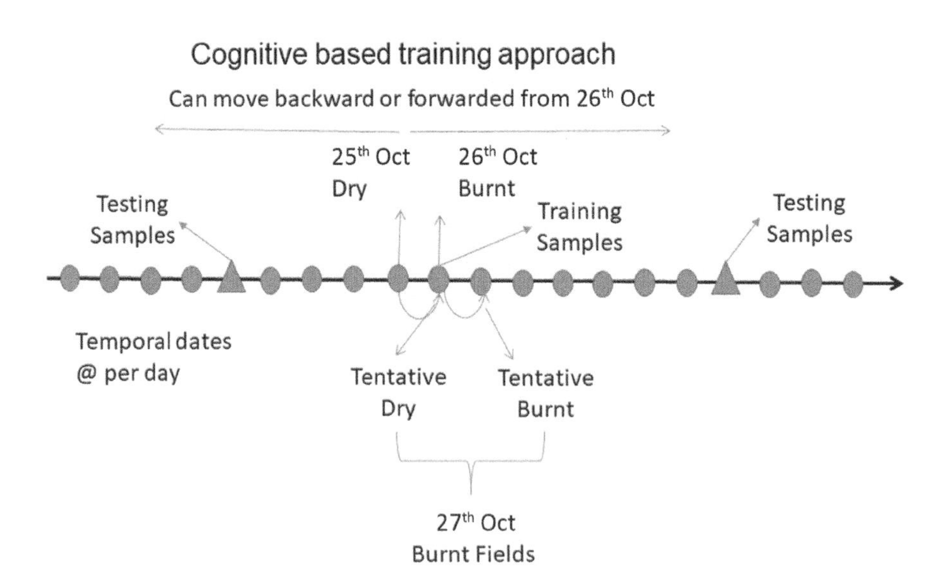

FIGURE 4.3 Example of a cognitive-based training approach.

been applied to the 27th October image to obtain tentative dry stubble from the 26th October image and tentative burnt stubble from the 27th October image, respectively. From these tentative outputs, training samples have been generated having maximum dry stubble converted to maximum dry stubble in 26th and maximum paddy burnt stubble in 27th October output images. At the location of maximum dry stubble converted to maximum burnt stubble in 26th and 27th October output images, temporal indices were generated, and at the same location, training data was generated to extract stubble burnt fields on the 27th October image. Further temporal dates for images will be processed in the same step for mapping stubble burnt fields.

4.6 SPECIFIC CLASS MAPPING APPLICATIONS

There are various applications of specific-class mapping, like specific crop acreage mapping, specific crop monitoring, specific crop insurance, and many more. In this section, specific crop acreage mapping, monitoring, and insurance have been mentioned (Joel et al., 2017).

Crop Acreage Mapping

Number of times, at farmer level, the agriculture land-area is not defined in acres however using local units of area. The acreage is another way to assess farm size. Acreage information of a crop is very important and it's first-level information. From acreage information, further secondary information can be generated, like yield estimation, diseased fields, flood affected crop fields, and so forth. As per size of fields in different countries, family status has been defined. Like in the United States Department of Agriculture (USDA), 231 acres are categorized as small land-areas; 1,421 acres, as large farm land areas; and 2,086 acres, as very large farm lands. It may come as a surprise to learn that small family farms account for 88% of all farms in the United States. Mapping specific crop acreage information from remote sensing data is a challenging task due to spectral overlap between various crops/vegetation present in an area. This problem can be solved while using temporal remote sensing images using mono/dual/multi-sensor data-sets. Further, using temporal remote sensing data itself is not the solution however, appropriate machine leaning models need to be selected to map single crop fields of interest. Moreover, spectral dimensionality of remote sensing data need to be reduced with methods like indices while retaining temporal crop information to incorporate phenological information of specific crops.

Specific Crop Monitoring

Crop monitoring activity by farmers, as well as by local government bodies, is routine within crop sessional activity to learn crop health, as well as the type of crops sown in an area. Generally, while monitoring crops, a farmer walks through crop fields to look for problems associated with crop. The monitoring of crops helps make farming chores easier, as well as providing the best user experience and produce a better output. Crop growth and performance are closely monitored during the developmental stages, which is an important component of agricultural management. The monitoring of crops helps the farmer in making decisions at the right time, so that crops result in the maximum yield at the harvesting of the season. Crops are frequently hampered by stress factors that prevent them from developing at their full potential. Crop health management employs current information and communication technologies, such as IoT-enabled sensors or drones to monitor crop health, as well as crucial environmental elements, such as moisture levels. The type of crops grown in a given location, the state of the crop, and the yield may all be examined using remote sensing. Extracting crop-stage information at some interval through remote sensing images provides crop status at an interval, as well as crop condition and growth progression in the fields. But finer, temporal resolution remote sensing data can be used to monitor specific crop stages. Presently, planetscope dove satellites datasets, available on daily basis, as well as UAV images, can also be acquired any time for crop-monitoring studies.

Specific Crop Insurance

Agricultural producers acquire crop insurance, which is subsidised by the federal government. Natural disasters such as deep freezes, floods, droughts, excessive wetness, sickness, hot weather and other natural calamities are covered. Crop insurance is divided into two categories: MPCI (multiple-peril crop insurance) and CHI (crop-hail insurance). These include natural-disaster events like destruction due to weather such as hail storms, night-time frost, damaging strong winds, crop disease, drought as an event of prolonged shortages in the water supply, anthropogenic fire, natural floods, and crop production damage due to deadly bugs.

In the areas where the occurrence of hail common, farmers generally apply for insurance of their crops against damage of the crop due to hail. Private insurers sell these plans, which are controlled by state insurance departments, and these programs are not part of government programs. As a supplement to MPCI, many farmers get crop hail insurance. Crop hail insurance typically has a very low premium or sometimes there is no premium. As the occurrence of hail is in localised areas, hail destroys a small portion of agricultural land having crops while leaving other crops unharmed. Due to the limitations of hail insurance, it is much less popular. Crop hail insurance can be applied any time during the growing session of the crop.

A second type of crop insurance is related to revenue insurance for farmers. The advantage of revenue insurance is that, when crop yield is less, it supports farmers. The calculation of revenue insurance is based on current-year revenue-loss, compared to last-year earnings. The overall revenue insurance benefit to farmers is to safeguard their earning against instability in crop prices.

In the upcoming year, crop insurance has been given a boost by the government with various products in the crop-insurance scheme. Crop-insurance experts can help farmers to decide which crop-insurance policy is beneficial.

Crop insurance mainly can be divided into two categories of insurance. The first one is yield-dependent coverage, which means that, if the yield is less, the insurance claim will be paid. The second insurance type is the revenue-plan type, which gives assurances based on low yield and low crop-price. Drought, excessive precipitation, hail, wind, frost, insects, and illness are all common natural factors that are generally covered. Price variations can also be accommodated. Pesticide drift, fire, negligence, failure to follow basic farming practices, and other causes of property loss are not covered. The guarantee is determined by multiplying the average output by the crop coverage-level chosen by the grower. If the harvested and appraised produce is less than the stipulated amount, an indemnity or loss payment may be owed.

The use of temporal remote-sensing data in conjunction with a machine-learning algorithm provides a powerful solution for mapping damage, diseased, or low-yield crop fields and for providing insurance in the event of a loss.

4.7 SUMMARY

This chapter discussed about the training-parameter concepts that should be considered in statistical or fuzzy supervised-classifiers. This chapter began with a discussion of manually collecting training samples, as well as region growing method for the same. The heterogeneity of training samples collected from images in pixel forms affects classification results, because classification results are not homogeneous. As a result, in this chapter, a training approach based on individual training-samples had been discussed. Further, this chapter has also highlighted about the training-sample size, mentioning them with limited samples collected from the ground as well as to extend training samples using statistical techniques. The final section had covered the application of single classe with examples in the agriculture crop sector. The fifth chapter discusses the role of specific machine-learning models in specific-class mapping.

BIBLIOGRAPHY

Abdi, Asad (2016). Cognitive science, *Why Cognitive Science*? https://doi.org/10.13140/RG.2.2.34708.71042.

Chaturvedi, A., R. Khanna & V. Kumar (2016). An analysis of region growing image segmentation schemes, *International Journal of Emerging Trends & Technology in Computer Science*, 34(1):46–51, https://doi.org/10.14445/22312803/IJCTT-V34P108.

Diamant, E. (2014). Cognitive image processing: The time is right to recognize that the world does not rest more on turtles and elephants, *arXiv 1411.0054.*

Giorgi, Giacomo De, William Gui Woolston & Michele Pellizzari (2012). Class size and class heterogeneity, *Journal of the European Economic Association*, 10(4):795–830.

Maxwell, Aaron E., Timothy A. Warner & Fang Fang (2018). Implementation of machine-learning classification in remote sensing: An applied review, *International Journal of Remote Sensing*, 39(9):2784–2817, https://doi.org/10.1080/01431161.2018.1433343.

Mubarak, D. M. N., M. M. Sathik, S. Z. Beevi & K. Revathy (2012). A hybrid region growing algorithm for medical image segmentation, *International Journal of Computer Science & Information Technology (IJCSIT)*, 4(3), https://doi.org/10.5121/ijcsit.2012.4306.

Roh Y., G. Heo and S. E. Whang (2021). A Survey on Data Collection for Machine Learning: A Big Data – AI Integration Perspective, *IEEE Transactions on Knowledge and Data Engineering*, 33(4):1328–1347, doi: 10.1109/TKDE.2019.2946162.

Salas, Eric Ariel L., Sakthi Kumaran Subburayalu, Brian Slater, Rucha Dave, Parshva Parekh, Kaiguang Zhao & Bimal Bhattacharya (2021). Assessing the effectiveness of ground truth data to capture landscape variability from an agricultural region using Gaussian

simulation and geostatistical techniques, *Heliyon*, 7(7):ISSN 2405–8440, https://doi.org/10.1016/j.heliyon.2021.e07439.
Silva, Joel, Fernando Bação & Mário Caetano (2017). Specific land cover class mapping by semi-supervised weighted support vector machines, *Remote Sensing*, 9(2):181, https://doi.org/10.3390/rs9020181.
Singhal, Mragank, Ashish Payal & Anil Kumar (2021). Procreation of training data using cognitive science in temporal data processing for burnt paddy fields mapping, *Remote Sensing Applications: Society and Environment*, 22:100516, ISSN 2352–9385, https://doi.org/10.1016/j.rsase.2021.100516.
Taherdoost, Hamed (2016). Sampling methods in research methodology; how to choose a sampling technique for research, *International Journal of Academic Research in Management (IJARM)*, 5(2):18–27, ISSN: 2296–1747.
Verwey, W. B., C. H. Shea & D. L. Wright (2015). A cognitive framework for explaining serial processing and sequence execution strategies, *Psychonomic Bulletin & Review*, 22:54–77, https://doi.org/10.3758/s13423-014-0773-4.

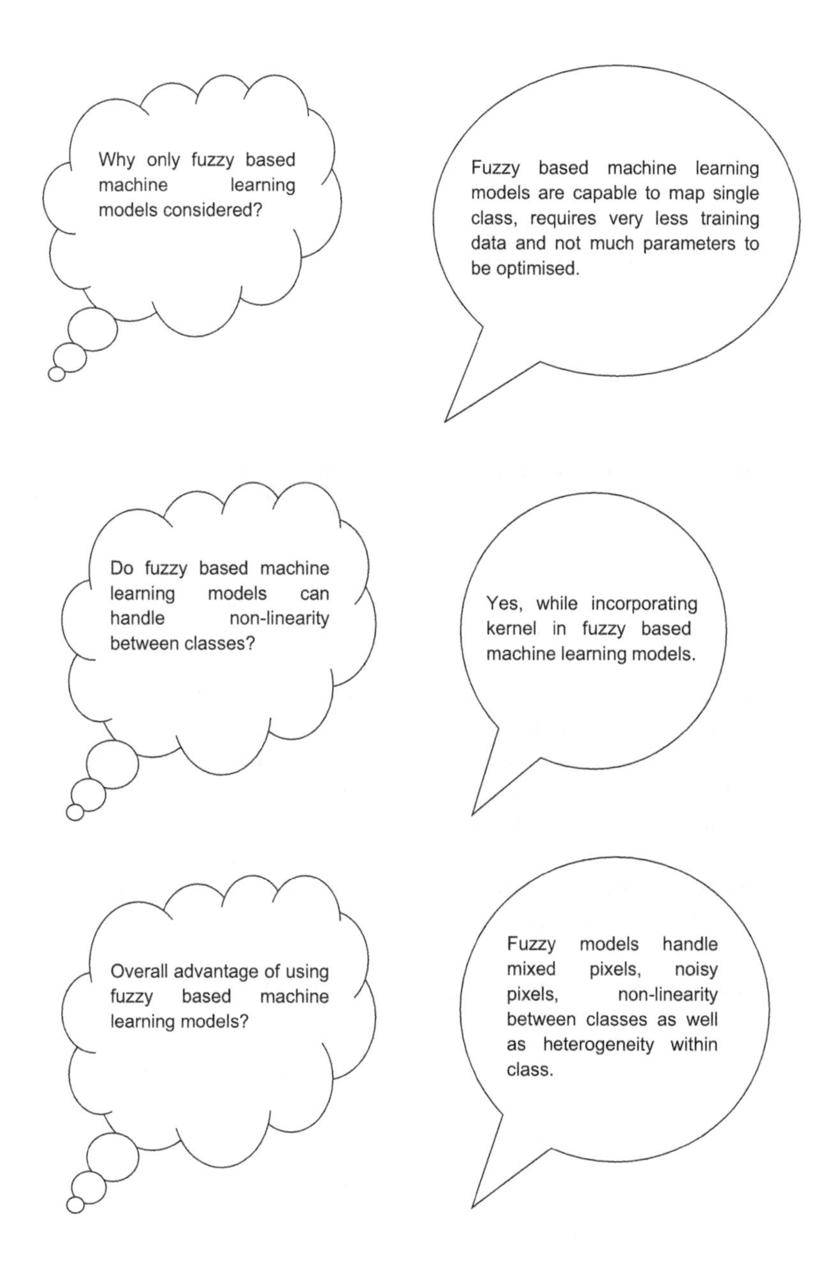

See Challenges as Opportunities for New Afresh Beginning. . . .

5 Machine-Learning Models for Specific-Class Mapping

5.1 INTRODUCTION

There are numerous machine-learning models available for use in a variety of applications. Machine-learning classifiers can fall into statistical categories such as *c*-Means, maximum-likelihood classifiers, and decision-tree categories, such as random forest and Classification and Regression Tree (CART). Fuzzy *c*-Means (FCM), Possibilistic *c*-Means (PCM), Noise Clustering (NC), and Modified Possibilistic *c*-Means (MPCM) are examples of fuzzy-logic algorithms. The machine-learning model should be chosen in such a way that it can deal with mixed pixels, non-linearity between classes, and noisy pixels. Machine-learning models should be able to map a single class of interest, which has a wide range of applications, where only one class of interest needs to be mapped from remote-sensing data. Machine-learning models should also be able to deal with heterogeneity within classes, which is caused by the fact that training data for each class is not homogeneous. So, in this section, fuzzy classifiers are discussed, which addresses the raised concerns.

In this chapter, mathematical formulas and algorithms of the fuzzy-based algorithms have been explained. The description starts with Fuzzy *c*-Means (FCM), moves on to Possibilistic *c*-Means (PCM) and Noise Clustering, and then Modified Possibilistic *c*-Means (MPCM) is described.

5.2 FUZZY SET-THEORY-BASED ALGORITHMS

Clustering is essentially an unsupervised classification technique in which comparable data points within clusters are grouped together while different data points are grouped in a distinct cluster. Clustering techniques have a wide range of applications, including data mining, image processing, pattern recognition, segmentation, and so on. They are partitional and hierarchical in nature. In partitional clustering, data is divided into a specified numbers of clusters; whereas, in hierarchical clustering, a dendorum-type structure of data is built (Soman, Diwakar, and Ajay, 2006). These methods can be hard or soft output according to the requirements (Dave, 1991). Every sample is assigned to a single cluster using the hard-clustering algorithm. This principle is expanded in the fuzzy-based clustering method, where each data item is associated with all of the clusters and provides a measure of belongingness or membership values. The clustering method should find an optimal or nearly ideal data partition. The pace of convergence is affected by several characteristics, including starting-step variance and global step-size variance (Babu and Murty, 1994).

DOI: 10.1201/9781003373216-5

As a result, many unsupervised algorithms may be classified as optimisation problems, which implies that they optimise their parameters through repetition. Hard clustering optimises only the cluster centres, whereas soft classification optimises both the cluster centres and the membership values of the data points. In fuzzy-based algorithms, the resulting classes are not 'hard' or 'crisp' any longer however become 'fuzzy'; which means any pixel can belong to two or more than two classes simultaneously and each pixel will have a different 'membership value' for each class (Mather and Tso, 2016). According to Bezdek (1981), fuzzy-based algorithms perform superiorly in comparison to their corresponding hard variants, as they do not generally stick to local minima. Also, the fuzzy model does not follow probability rules, due to which these fuzzy models can be used to do mapping of specific classes.

5.3 FUZZY c-MEANS (FCM) ALGORITHM

FCM is basically a clustering algorithm which uses membership values for defining elements belonging to a class (Bezdek, 1981). It is an iterative clustering technique. It follows a 'hyper-line constraint', which says that the summation of membership values of all classes present in a pixel should be one (Krishnapuram and Keller, 1996). The objective function of FCM is given in Eq. (5.1), in which it needs to be minimised to attain the clustering criteria:

$$J_{FCM}(U,V)=\sum_{i=1}^{N}\sum_{j=1}^{c}(\mu_{ij})^{m}\,d_{ij}^{\,2} \tag{5.1}$$

with constraints as:

For all i $\quad \sum_{j=1}^{c}\mu_{ij}=1$

For all j $\quad \sum_{i=1}^{N}\mu_{ij}>0$

For all i, j $\quad 0 \le \mu_{ij} \le 1$

Here, d_{ij}^2 is the distance between v_j and x_i in feature space and is given by Eq. (5.2)

$$d_{ij}^2 = \| x_i - v_j \|^2 = (x_i - v_j)^T A (x_i - v_j) \tag{5.2}$$

'μ_{ij}' represents the membership value of an element or pixel 'i' to belong to class 'j'; 'N' represents the total number of pixels in an image; 'v_j' represents cluster mean as its centre of class 'j'; 'x_i' represents the unknown vector at 'i'; 'A' indicates the Euclidean norm used; 'm' represents membership weight constant ($1<m<\infty$) which governs the degree of fuzziness (if m value is towards 1, J_{FCM} value goes towards hard, and as m tends to ∞ it goes towards soft). FCM algorithm is an iterative method where the pixels are partitioned and given different membership values for each class. This partitioning is achieved by optimising Eq. (5.1). Cluster centre is iteratively calculated and updated using Eq. (5.3) (Mather and Tso, 2016).

$$v_j = \frac{\sum_{i=1}^{N} \mu_{ij}^{m} . x_i}{\sum_{i=1}^{N} \mu_{ij}^{m}} \tag{5.3}$$

The class membership (μ_{ij}) is then calculated using Eq. (5.4)

$$\mu_{ij} = \left[\sum_{k=1}^{c} \left(\frac{d_{ij}^{2}}{d_{ik}^{2}} \right)^{1/m-1} \right]^{-1} \tag{5.4}$$

where,

$$d_{ik}^{2} = \sum_{j=1}^{c} d_{ij}^{2}$$

Membership values are the shares of all classes present in a pixel, which can vary from 0 to 1 (Mather and Tso, 2016).

5.4 POSSIBILISTIC *c*-MEANS CLASSIFICATION

Fuzzy-logic theory provides a major advantage in classification, as it permits the natural explanation of the problem in linguistic terms and not in terms of just the relationships among the specific numerical quantities (Venkateswaran, Vijaya, and Saravanan, 2013).

Bezdek, Ehrlich, and Full (1984) developed the first fundamental fuzzy algorithm, FCM, to deal with the problem of mixed pixels. According to them, the objective of FCM is to represent the data point's similarity to the cluster centre and not just represent the cluster's centre properties. Now, this technique is used for remote sensing images for its ability to handle the uncertainty in the pixel, so it can preserve more information than the traditional hard classifiers. According to Krishnapuram and Keller (1996), FCM provides relevant information regarding the degree of belongingness of all the sample points having dissimilar cluster means. They stressed a particular limitation of FCM that, in a pixel the sum of membership values should be equal to one. Furthermore, Krishnapuram and Keller (1993) and Wu and Zhou (2006), found that FCM is more sensitive to outliers and noise present in the image, as it represents the degree of sharing in place of the degree of typicality. They proposed an algorithm based on the concept of possibility (Zadeh, 1978), with an improvement in the objective function, and is called Possibilistic *c*-Mean (PCM).The hyper-line constraint for membership values is not followed, as a membership value of a feature vector belonging to one class has no relation with the feature vector of other classes.

Foody (2000), observed that PCM works well for untrained classes, as well as giving low Root Mean Square Error (RMSE) in the case of untrained classes.

In many applications, the interest is only in a particular class, rather than all the classes present in the remotely sensed image. Here, the classifier can be trained by the samples of only one class, with very few samples from the other classes (Sun et al., 2008). Their research focused on active one-class categorization using an active learning-technique and support-vector data description. Crop categorisation is another example of a situation in which single-class mapping information is required. Upadhyay, Kumar, Roy, Ghosh, and Ian (2012) used the Possibilistic *c*-Means (PCM) algorithm to extract a single class, as in this algorithm, the membership value of a class in a pixel is not dependent on the membership values of other classes present in the same pixel. Because it does not follow the probability rule, the sum of the membership values of the classes in a pixel may not equal to unity. Water was extracted from AWiFS sensor data of the Resourcesat-1 satellite. The classification accuracy attained was in the range of 84–99%.

The Possibilistic *c*-Mean algorithm is the extended version of the Fuzzy *c*-Mean algorithm. In this algorithm, the elements belonging to a class were given high membership, and the elements not belonging to a class were given small membership value (Krishnapuram and Keller, 1993).

According to the possibilistic method, the membership value of a pixel in a class displays the typicality of that pixel in the class, or the likelihood of that pixel lying inside that class (Krishnapuram and Keller, 1993; Wu and Zhou, 2008).

In PCM, the objective function of FCM is slightly modified mathematically. PCM relaxes the hyper-line constraint on membership values. Memberships that represent the typicality can be formed by loosening its limitation. Because the noisy points are less typical, it aids in noise reduction and thus improves the results (Krishnapuram and Keller, 1993; Chawla, 2010).The objective function followed by PCM is (Eq. 5.5):

$$J_{PCM}(U,V)=\sum_{i=1}^{N}\sum_{j=1}^{c}\left(\mu_{ij}\right)^{m} d_{ij}^{2}+\sum_{j=1}^{c}\eta_{j}\sum_{i=1}^{N}\left(1-\mu_{ij}\right)^{m} \tag{5.5}$$

The constraints to PCM objective function are given in Eq. (5.6):

$$\begin{aligned}&\text{For all } i^{\max}\ \mu_{ij}>0;\ \forall j\\&\text{For all } j \sum_{i=1}^{N}\mu_{ij}>0\\&\text{For all } i\ 0\le\mu_{ij}\le 1\end{aligned} \tag{5.6}$$

Here $d_{ij}^{2} = \| x_i - v_j \|^2$ is the distance between v_j and x_i in feature space; 'μ_{ij}' represents the membership value of unknown vector '*i*' for belonging to class '*j*'; '*N*' is the number of pixels in an image; 'c' represents number of classes; 'v_j' represents the cluster centre calculated for class '*j*'; 'x_i' represents the feature vector for pixel '*i*'; '*A*' represents the Euclidean norm used in this study; 'm' represents membership weight constant ($1<m<\infty$) and controls the degree of fuzziness.

'η_j', called the 'Regularisation Factor', depends on the cluster average size and shape of the cluster j, and it is calculated as Eq. (5.7):

$$\eta_j = K\frac{\sum_{i=1}^{N}\left(\mu_{ij}\right)^m d_{ij}^{\,2}}{\sum_{i=1}^{N}\left(\mu_{ij}\right)^m} \tag{5.7}$$

'K' denotes a constant here and its value is generally 1. Then, the class membership values (μ_{ij}) are calculated using Eq. (5.8):

$$\mu_{ij} = \frac{1}{1+\left(\frac{d_{ij}^{\,2}}{\eta_j}\right)^{1/m-1}} \tag{5.8}$$

Eq. (5.4) shows that that number of classes needed to be extracted decide the membership values that are generated for the pixels, as $d_{ik}^{\,2} = \sum_{j=1}^{c} d_{ij}^{\,2}$. So, while extracting only a single class from an image, then $d_{ik}^2 = d_{ij}^2$ while membership values (μ_{ij}) of all pixels belonging to class j become one. This indicates that each pixel will belong to the same class, which is not required. Therefore, in PCM, 'η_j' is calculated as $\eta_j = K\frac{\sum_{i=1}^{N}\left(\mu_{ij}\right)^m}{N}$ and class memberships are found by using Eq. (5.8).

Although PCM performs very well as a data-clustering algorithm, it suffers from certain drawbacks. Krishnapuram and Keller (1996), explained that the PCM algorithm has an undesirable tendency of producing coincident clusters and is sensitive to good initialisation. It is due to the fact that the typicality matrix has rows and columns which are independent to each other (Wu & Zhou, 2008). Another limitation of PCM is that it ignores that membership value which makes the class centroid close to data points, although it minimises the impact of noise. Therefore, a modified variant of PCM was proposed for fitting the clusters that are close to each other. This modified version was named as a modified possibilistic c-mean (MPCM) algorithm (Wu and Zhou, 2008). Before explaining the MPCM fuzzy-model let's understand the Noise Clustering (NC) classifier.

5.5 NOISE CLUSTERING

Outliers and the noise present in the data have always been a hurdle in effective clustering. If data contains some noise, it produces a certain bias to the clustering algorithm and hence leads to unrealistic clusters. Therefore, prior knowledge of clusters and noise present in the dataset are the two main factors that affect the performance of a clustering algorithm (Harikumar, 2014). According to Bezdek (1981) the number of clusters that the data can be divided into can be found using certain measures, to some level of accuracy. Krishnapuram and Freg (1992), found good results using a compatible cluster-merging algorithm. Dave (1991), also concluded that noise clustering gave the best performance.

As FCM has been found to be sensitive towards outliers and noise, later, a Noise Clustering (NC) algorithm was also suggested to overcome these constraints of FCM. In NC, a minimum constant distance is calculated from all cluster centres, and this is referred to as noise distance. Later, Davé and Sen (2002) discovered that treating noise as a constant makes the algorithm stiff; thus, it was optimised to allow varying noise distances to be calculated for different feature vectors. It was also proved mathematically by Davé and Krishnapuram (1997) that the Noise Clustering algorithm is actually a simplification form, and FCM and PCM algorithms are the extraordinary forms of NC. The main disadvantage of the PCM algorithm is that its performance is subject to good initialisation. The FCM algorithm can provide good cluster initialisation and so it is utilised for estimating the cluster centres for the PCM algorithm. So, the noise-clustering algorithm provides a combination of the mode seeking ability of the PCM algorithm and the FCM algorithm's partitioning ability, and that is why it gives a relatively much better performance, according to Davé and Krishnapuram (1997).

Figure 5.1, here, depicts the sample data-set, where (a) shows the noisy data and (b) shows the noise-clustering result. The '0' and '◊' depicts two dusters and '⊔' depicts the noisy data. Figure (b) shows the clusters formed.

As already stated, noisy data have always posed a problem in effective clustering. Noise present in the data causes the clustering algorithm to become biased, and hence, the clusters produced are unrealistic in nature.

Davé and Krishnapuram (1997) introduced a noise-clustering model, which was designed as an alternate way to deal with the sensitivity that FCM posed to noisy data or outliers. It was also emphasised that all noisy data should be kept in a separate cluster. The noise-clustering algorithm was proposed following fuzzy form, as this noise-clustering model provides membership values. Through the noise-clustering model, while creating clusters of interest, this fuzzy model also creates one extra cluster for noisy elements. In the noise-clustering model, equal distance of noisy elements with other clusters of interest is considered to create noisy clusters, as it takes care that each data element has equal probability for belonging to noisy clusters (Dave, 1991).

The noise-distance parameter (δ) of the noise-clustering algorithm is the distance between clusters of interest and noisy clusters. It is the most vital parameter

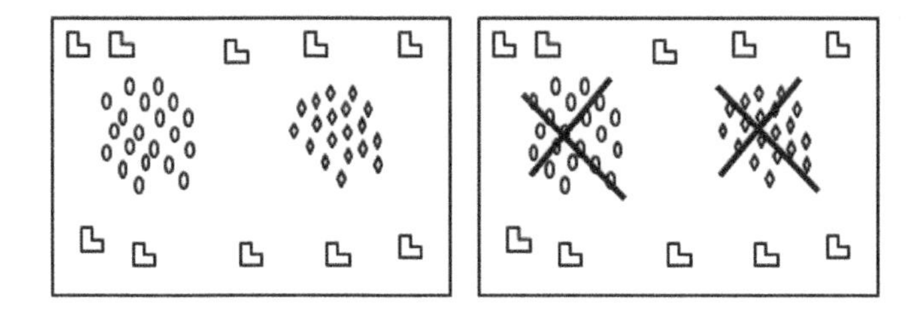

a. Data having some noise *b. Noise clusters formed by NC algorithm*

FIGURE 5.1 Noise-Clustering algorithm, Dave (1993).

which directly influences and decides the way a noise clustering algorithm performs (Dave, 1991). The noise-distance parameter (δ) distance is actually governed by the data points but could be estimated using Eq. (5.9) (Dave, 1991):

$$\delta^2 = \lambda \left[\frac{\sum_{i=1}^{N} \sum_{j=1}^{c} D(\vec{x}_i, \vec{v}_j)}{N_c} \right] \tag{5.9}$$

Here, all the data points which lie beyond a threshold form the cluster centres; that is, those having higher distance than the noise distance are clubbed in noise cluster. For Noise Clustering, the constraint on membership is: $\sum_{j=1}^{c+1} \mu_{ij} \leq 1,\ 1 \leq i \leq N$

The range of the sum of membership values for a given element belonging to all clusters can be equal to one or less than one, as well. So, a number of times, it has been observed that noise points generally have small values of membership, as well (Davé and Krishnapuram, 1997). Also, from Eq. (5.10), as there is no mean for noisy clusters, noisy elements will not be affected with noisy-cluster centres.

The objective function of the noise-clustering algorithm is given in Eq. (5.10):

$$J_{NC}(U,V) = \sum_{i=1}^{N} \sum_{i=1}^{c} (\mu_{ij})^m D(\vec{x}_i, \vec{v}_j) + \sum_{i=1}^{N} (\mu_{ic+1})^m \delta \tag{5.10}$$

The first term of Eq. (5.10) is equivalent to FCM's objective function. The second part, as a regularising function in Eq. (5.10), is for generating noisy cluster (c+1), as well as for providing membership values for noisy elements.

In Eq. (5.10), 'c' is the total number of clusters; 'N' is the total number of pixels present in an image; 'm' controls fuzziness in output and is called the fuzzification factor; 'μ_{ij}' is called membership values of an element or pixel 'i' belonging to class 'j'; 'δ' is called noise distance; '$u_{i,c+1}$' is called noise class membership values; '$\vec{x}_j$' is the cluster centre of 'j^{th}'class; 'D' is called the square of Euclidean distance between '$\vec{x}_i$' and '$\vec{v}_j$', '$\vec{x}_i$' is an unknown vector element at a pixel 'i'. Eq. (5.11), Eq. (5.12), and Eq. (5.13) have been generated from Eq. (5.10), which provides the membership value for a pixel belonging to a cluster, noisy element/pixel belonging to a noisy cluster, and fuzzy-cluster centre, respectively.

$$\mu_{ij} = \left[\sum_{k=1}^{c} \left(\frac{D(\vec{x}_i, \vec{v}_j)}{D(\vec{x}_i, \vec{v}_k)} \right)^{\frac{1}{m-1}} + \left(\frac{D(\vec{x}_i, \vec{v}_j)}{\delta} \right)^{\frac{1}{m-1}} \right]^{-1},\ 1 \leq i \leq c \tag{5.11}$$

$$\mu_{i,c+1} = \left[\sum_{k=1}^{c} \left(\frac{\delta}{D(\vec{x}_i, \vec{v}_j)} \right)^{\frac{2}{m-1}} + 1 \right]^{-1} \tag{5.12}$$

$$\vec{v}_j = \frac{\sum_{i=1}^{N} (\mu_{ij})^m \vec{x}_i}{\sum_{i=1}^{N} (\mu_{ij})^m}, \ 1 \leq i \leq c \tag{5.13}$$

For Noise Clustering, the constraint on membership value is shown in Eq. (5.14):

$$\sum_{j=1}^{c+1} \mu_{ij} \leq 1, \ 1 \leq i \leq N \tag{5.14}$$

The key advantage of the NC algorithm is its ability to handle noise and outliers. This capability of NC in dealing with the fuzziness of the pixels in the image and allocating a separate class to all the noisy data can produce more realistic classification results (Richards and Jia, 2006). Noise classifiers can also handle the untrained classes (the classes which are there in the image but the classifier has not been trained for them) very well. For the untrained/unclassified elements/pixels also, the noise-clustering model will assign them to a noisy class. In the next section, mathematical details of the MPCM classifier have been mentioned.

5.6 MODIFIED POSSIBILISTIC *c*-MEANS (MPCM) ALGORITHM

MPCM (Modified Possibilistic *c*-Mean) fuzzy-classification model was developed by Li, Huang, and Li (2003) to remove the restrictions in FCM and PCM algorithms. As PCM is affected due to concurrent clusters issues, so a modified version of PCM was suggested, called MPCM. The MPCM classifier has the capability of resisting noise and avoids insignificant solution, with a fast cluster-formation ability (Li, Huang, and Li, 2003). MPCM fuzzy algorithm follows the following steps:

a) Mean value is calculated for each class.
b) Value of the degree of fuzziness (m) >1, is assigned.
c) Regularisation parameter η_j is calculated using Eq.(5.7).
d) Noise minimiser parameter λi is computed. λi is used to minimise the effect of outliers and noise. It was introduced for each training sample and objective function of PCM. It is computed from membership matrix (μ_{ij})using Eq. (5.15):

$$\lambda_i = \mu_{ij} \log(\mu_{ij}) \tag{5.15}$$

Membership matrix (μ_{ij}) is computed using Eq. (5.16):

$$\mu_{ij} = e^{\left(\frac{-d_{ij}^2}{\eta_i}\right)} \tag{5.16}$$

Assign the final class values to every pixel.

The objective function of MPCM is given in Eq. (5.17):

$$J_{MPCM}(U,V)=\sum_{j=1}^{C}\sum_{i=1}^{N}\mu_{ij}\left\|x_i-v_j\right\|^2+\eta_i\sum_{i=1}^{N}(\mu_{ij}\log(\mu_{ij})-\mu_{ij}) \quad (5.17)$$

Here, $\lambda_i > 0$.

'd_{ij}^2' is the distance between x_i and v_j in feature space and is given by Eq. (5.18):

$$d_{ij}^2 = \| x_i - v_j \|^2 \quad (5.18)$$

η_i represents the distribution parameter computed by using Eq. (5.19):

$$\eta_i = \frac{\sum_{i=1}^{N}(\mu_{ij})^m{}_{FCM}\, d_{ij}^2}{\sum_{i=1}^{N}(\mu_{ij})^m{}_{FCM}} \quad (5.19)$$

'v_j' is the cluster centre calculated as Eq. (5.20):

$$v_j = \frac{\sum_{i=1}^{N}\mu_{ij}x_i}{\sum_{i=1}^{N}\mu_{ij}} \quad (5.20)$$

From Eq. (5.19), it was noticed that, while mapping a single class of interest, the distribution parameter 'η_i' is not a dependent of 'm' fuzziness factor parameter. As an outcome of this, the complete objective function of the MPCM algorithm at any stage is no longer a dependent of 'm' fuzziness factor parameter. So, the MPCM fuzzy algorithm, while mapping a single class of interest, becomes independent of parameters to be optimised.

Secondly, all fuzzy classifiers mentioned in this chapter can be used unsupervised, called clustering as well as supervised classification mode.

5.7 SUMMARY

In this chapter, dedicated, fuzzy machine-learning models capable of mapping a single class of interest have been discussed. Initially, a basic, fuzzy FCM classifier was discussed, which is incapable of mapping a single class of interest due to a probability limitation. Following FCM, several fuzzy classifiers capable of mapping a single class of interest such as PCM, NC, and MPCM are discussed. These three classifiers can manage noise and hence they can map a single class of interest. MPCM working for single-class mapping does not require any parameter to be optimised. Next, Chapter 6 is on learning-based classifiers in which an artificial neural network as base classifier with more focus on deep, learning-based classifiers like CNN, RNN, and its versions are discussed.

BIBLIOGRAPHY

Babu, G. P. & Murty, M. N. (1994). Clustering with evolution strategies. *Pattern Recognition*, 27, 321–329. http://doi.org/10.1016/0031-3203(94)90063-9.

Bezdek, J. C. (1981). *Pattern Recognition with Fuzzy Objective Function Algorithms*. Springer New York, NY. http://doi.org/10.1007/978-1-4757-0450-1.

Bezdek, J. C., Ehrlich, R. & Full, W. (1984). FCM: The fuzzy c-means clustering algorithm. *Computers & Geosciences*, 10, 191–203. http://doi.org/10.1016/0098-3004(84)90020-7.

Chawla, S. (2010). *Possibilistic-C-Means-Spatial Contexutal Information Based Sub-Pixel Classification Approach for Multi-spectral Data*. MSc. Thesis, University of Twente, Faculty of Geo-Information and Earth Observation (ITC), Enschede.

Dave, R. N. (1991). Characterization and detection of noise in clustering. *Pattern Recognition Letters*, 12, 657–664. http://doi.org/10.1016/0167-8655(91)90002-4

Davé, R. N. & Krishnapuram, R. (1997). Robust clustering methods: A unified view. *IEEE Transactions on Fuzzy Systems*, 5(2), 270–293. https://doi.org/10.1109/91.580801.

Davé, R. N. & Sen, S. (2002). Robust fuzzy clustering of relational data. *IEEE Transactions on Fuzzy Systems*, 10(6), 713–727 [p 2, 3, 4, 5]. https://doi.org/10.1109/TFUZZ.2002.805899.

Foody, G. (2000). Estimation of sub-pixel land cover composition in the presence of untrained classes. *Computers & Geosciences*, 26, 469–478. https://doi.org/10.1016/S0098-3004(99)00125-9.

Harikumar, A., Kumar, A., Stein, A., Raju, P. L. N. & Krishna Murthy, Y. V. N. (2015). An effective hybrid approach to remote-sensing image classification. *International Journal of Remote Sensing*, 36(11), 2767–2785.

Krishnapuram, R. & Freg, C.-P. (1992). Fitting an unknown number of lines and planes to image data through compatible cluster merging. *Pattern Recognition*, 25(4), 385–400. https://doi.org/10.1016/0031-3203(92)90087-y.

Krishnapuram, R. & Keller, J. M. (1993). A possibilistic approach to clustering. *IEEE Transactions on Fuzzy Systems*, 1(2), 98–110.

Krishnapuram, R. & Keller, J. M. (1996). The possibilistic *c*-means algorithm: Insights and recommendations. *IEEE Transactions on Fuzzy Systems*, 4, 385–393.

Li, K., Huang, H. & Li, K. (2003). A modified PCM clustering algorithm. *Proceedings of the 2003 International Conference on Machine Learning and Cybernetics (IEEE Cat. No.03EX693)*, vol. 2, 1174–1179, Xi'an, China IEEE. https://doi.org/10.1109/ICMLC.2003.1259663.

Mather, P. & Tso, B. (2016). *Classification Methods for Remotely Sensed Data*, 2nd ed. CRC Press, Boca Raton. http://doi.org/10.1201/b12554.

Richards, J. A. & Jia, X. (2006). *Remote Sensing Digital Image Analysis: An Introduction*. Berlin: Springer Verlag.

Soman, K. P., Diwakar, S. & Ajay, V. (2006). *Insight into Data Mining: Theory and Practice*. New Delhi: Prentice-Hall of India.

Sun, J. G., Liu, J. & Zhao, L. Y. (2008). Clustering algorithms research. *Journal of Software*, 19, 48–61. https://doi.org/10.3724/SP.J.1001.2008.00048.

Upadhyay, P., Kumar, A., Roy, P. S., Ghosh, S. K. & Ian, G. (2012). Effect on specific crop mapping using worldview-2 multispectral add-on bands: Soft classification approach. *Journal of Applied Remote Sensing*, 6(1), 063524. https://doi.org/10.1117/1.jrs.6.063524.

Venkateswaran, J. C., Vijaya, R. & Saravanan, A. M. (2013). A fuzzy based approach to classify remotely sensed images. *International Journal of Engineering and Technology (IJET)*, 5(3). ISSN: 0975–4024.

Wu, XH., Zhou, JJ. (2006). *An Improved Possibilistic C-Means Algorithm Based on Kernel Methods*. In: Yeung, DY., Kwok, J.T., Fred, A., Roli, F., de Ridder, D. (eds) Structural, Syntactic, and Statistical Pattern Recognition. SSPR /SPR 2006. Lecture Notes in Computer Science, vol 4109. Springer, Berlin, Heidelberg. https://doi.org/10.1007/11815921_86

Wu, X.-H. & Zhou, J.-J. (2008). Novel possibilistic fuzzy c-means clustering. *Tien Tzu Hsueh Pao/Acta Electronica Sinica*, 36, 1996–2000.

Zadeh, L. A. (1978). Fuzzy sets as a basis for a theory of possibility. *Fuzzy Sets and Systems*, 1(1), 3–28. ISSN 0165–0114.

How deep learning models are useful for temporal image processing?

For classification from single date or multi-temporal images, ID-CNN model is very effective for processing and classification.

Which deep learning models to be used for temporal image processing?

1D-CNN, RNN, LSTM models are very effective to processing temporal images, with minimum hardware resources requirement.

Spirituality Means. . . .

Life Transformation Journey. . . .

6 Learning-Based Algorithms for Specific-Class Mapping

6.1 INTRODUCTION

The learning-based models have been developed in different forms. Learning-based models can learn past knowledge, including cognitive capacities, by engaging, demonstrating information, and detecting new evidences. Researchers have been striving to develop computers with skills comparable to human beings since the dawn of the computer age (Michalski et al., 1983). Artificial Intelligence (AI) is a subset of machine intelligence that allows machines to learn and reason in the same way that humans do. Machine learning refers to the process through which a machine learns and models the learning process (Figure 6.1) (Michalski et al., 1983). Deep learning (DL) was established as a new subtopic within machine learning in 2006 (Figure 6.2) (Vargas et al., 2018). Basic models and components of deep learning are discussed in this section.

6.2 CONVOLUTIONAL NEURAL NETWORKS (CNN)

Walter Pitts and Warren McCulloch developed the first neural network model 1943. Initially, it was called McCulloh-Pitts neurons (Fogg, 2017). Artificial Intelligence (AI) was introduced in the 1950s by some pioneers of computer science. A crisp description of AI as given by Pelletier et al. (2019) is 'the effort to automate intellectual tasks normally formed by humans'. Machine learning is an application of AI, as well as subset of AI, while deep learning is a subset of machine learning. It is also known as a Machine Learning (ML) system, since it is learned by data rather than being manually coded. Many instances are presented to the model, allowing the system to develop certain rules and find some meaningful statistical

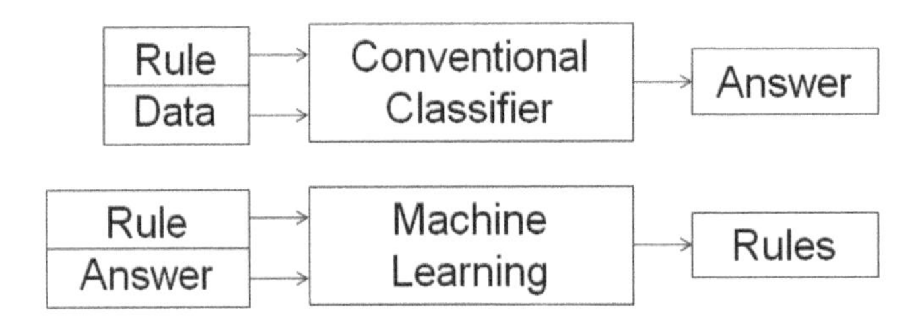

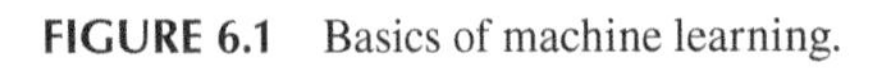
FIGURE 6.1 Basics of machine learning.

DOI: 10.1201/9781003373216-6

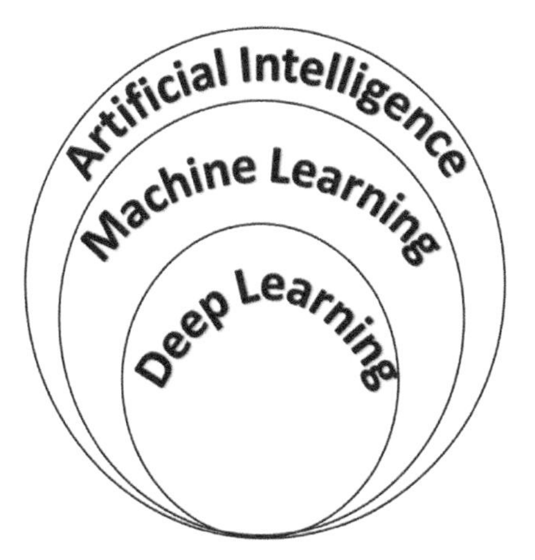

FIGURE 6.2 Deep learning as a subset of machine learning.

relationships in these examples. ML started gaining importance in the 1990s and soon but became very popular. The earliest deep-learning (DL) networks were developed in 1965 by V. G Lapa and Alexey Ivakhnenko. They developed these networks from the deep layers of feed-forward multilayer perceptrons by applying some statistical methods to each layer, searching for the best features, and propagating them in the network (Fogg, 2017). One of the basic models of deep learning is CNN (Convolutional Neural Network). CNNs are basically feed-forward networks, with architecture following the cell structure of animals' visual cortex, where every neuron responds to a respective field (Saha, 2018). CNN contains a pair or sometimes large number of convolution layers (Lecun et al., 2015). Convolution works like a kernel which convolves on an input data-set to extract useful features as information, while it removes redundant or unwanted information. The extracted features pass through a number of layers of convolution having an in-built activation function. At some intermediate intervals, there is pooling, while in the end, information is fed into fully connected layers (Figure 6.3). The role of pooling layers is to reduce feature size.

In recent times, CNNs have been extensively used for processing remote sensing images for various applications, like semantic segmentation, high-resolution image-classification, object detection, etc. (Maggiori et al., 2017). Various CNN models have also been applied for hyperspectral-image classification. Though, 1D-CNN performance was evaluated with respect to spectral information (Zhang et al., 2015), while two-dimensional CNNs (2D-CNN) were also evaluated with respect to spatial information (Liang and Li, 2016). While processing multi-temporal images for classification, the 1D-CNN model as well as the 2D CNN model can be useful, but still, extensive testing of 1D-CNN has not been done in the classification domain. There is great potential for the application of the 1D-CNN model for processing temporal remote-sensing data-sets.

RNNs are yet another category of DL networks that have been created to handle the sequential type of data. Due to this reason, RNNs are among the most studied

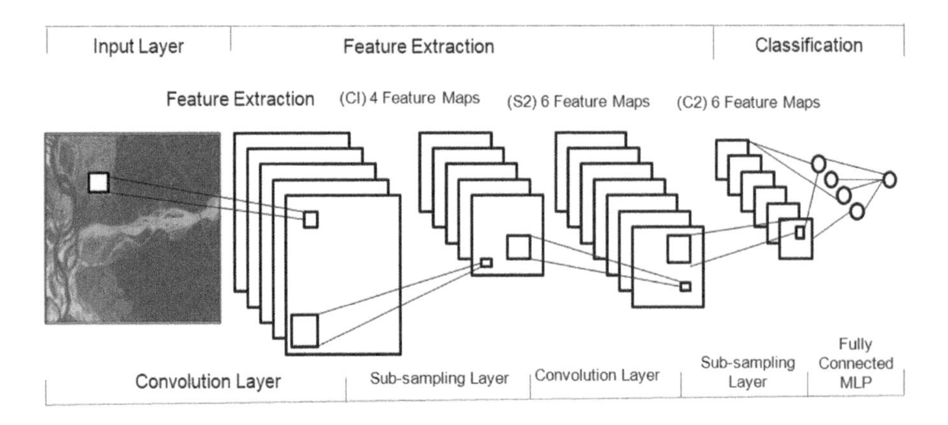

FIGURE 6.3 Architecture of CNN.

deep-learning (DL) architectures for the satellite-image time-series classification (Pelletier et al., 2019). RNN carries past information like human beings and its outputs are affected due to learning from the previous data. It takes any number of sets as input vectors and generates the equivalent number of sets as input vectors, which are influenced not only by the weights but also by the hidden-state vector. This signifies the context which is based on earlier input or output (Figure 6.4). Rubwurm and Korner (2017) and Ndikumana et al. (2018) used RNN networks for temporal vegetation modelling using medium-resolution satellite images. They tested two units of RNN; namely, LSTM (Long-short Term Memory) and GRUs (Gated Recurrent Units), where GRU showed less accuracy gain than LSTM. In these works, it was noticed that RNN performed better than other machine-learning algorithms like random forest and SVM. There is the potential to combine CNN and RNN approaches and test this hybrid strategy for categorising a single class of interest utilising multi-sensor temporal data.

The basic fundamental of learning algorithms is the mapping of input labels to the output labels. Here, the input layer is fed with a number of examples and the expected outputs. The input layer takes the input and gives some initial weight to it. In subsequent layers, the previous layers' output becomes input to the forward layer, and in all the layers, weights are generated. With the help of these weights, the output is predicted.

Subsequently, it is checked whether the predicted weight is similar to the expected output or not. If it is so, then it means that weights are correctly selected. However, this is rarely the case because the weights are initially chosen at random. As a result, a loss function is constructed, which gives the anticipated output's departure from the expected output. The variance is back-propagated, and the weights are then modified. The goal is to lower the loss function by adjusting the weights. This technique is conducted iteratively in order to reduce the loss function. The network is deemed to be trained when the loss function is minimised to a tolerable value. It is then fed test data in order to predict test results.

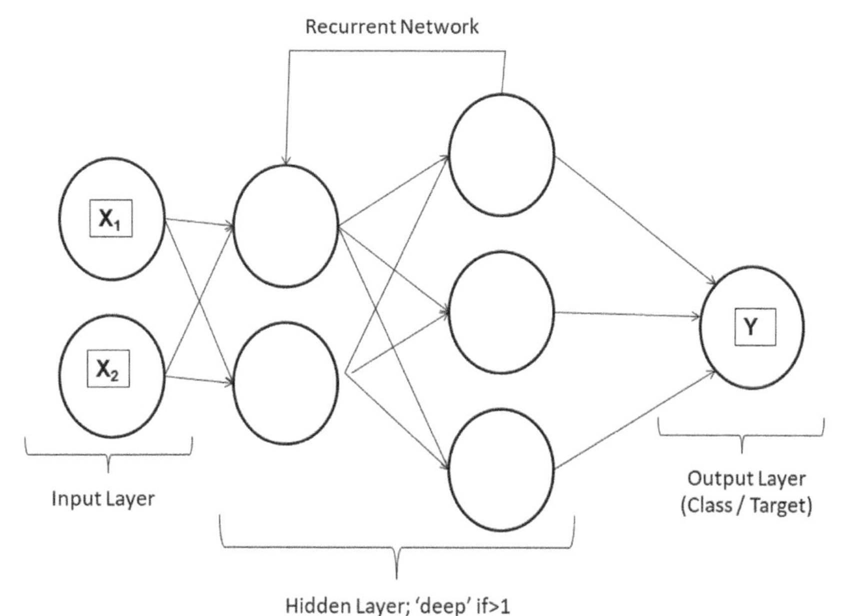

FIGURE 6.4 Structural design of Recurrent Neural Network (RNN).

Convolutional Neural Networks (CNN)

Deep learning helps to create computational architectures which consist of many layers that learn from the data representation using various layers of abstraction (LeCun et al., 2015). It consists of many convolutional layers along with the pooling layers which perform the extraction of features, as well as a data-reducing technique, respectively. In the last stage of the model, a fully connected layer is present with an output layer to produce classified output-results (Huang et al., 2019). CNN is comprised of several steps, which include transformation of features, fusion of information, feature extraction, and pattern identification, and it fits them into a deep architecture for optimisation and diagnosis (Yoo, 2015). In the contest of image processing, where the image is of a two-dimensional vector or can be equivalent to a matrix, it has been found that 2D-CNNs are the most suitable models for image processing. On the other hand, temporal signal processing, being a 1D (one dimensional) signal, can be easily done through 1D-CNNs. Time-varying signals are represented to be a 1D-vector or can be equivalent to a 1D-array or a 1D-matrix (Huang et al., 2019).

In the remote sensing domain, while considering temporal remote-sensing images, the temporal information can also be considered as a time-varying 1D signal. Representing temporal information from remote-sensing images as a time-varying 1-D signal and applying 1D-CNN models for processing have a lot of advantages. Firstly, it points towards time-varying data-processing being of the originality of the time-varying signal, which authenticates input data. Secondly, 1D-CNN uses 1D-convolution, which makes its structure much simpler, with very few parameters, in comparison to 2D-CNN. As a result, it necessitates relatively few computational hardware resources, as well as very little processing and time. Finally, owing to minimum number of parameters, there is little training data required during supervised processing (Huang et al., 2019; Kiranyaz et al., 2019).

1D-CNN Structure

The basic architecture of 1D-CNN is presented in Figure 6.5. The time-series, or temporal-image data is taken as input to the input layer; this data is processed by the first convolution layer, which extracts a large number of defined-feature layers. These convolution layers include kernels, as well as activation functions, to add non-linearity in the data. Later, one of the intermediate processes applied is called sub-sampling as the max or average pooling layer. The pooling layer's role in the CNN model is to reduce the spatial size of the image, as well as model parameters. These sub-sampling layers operate on every feature map independently. Lastly, there is the fully connected network, where classification is performed.

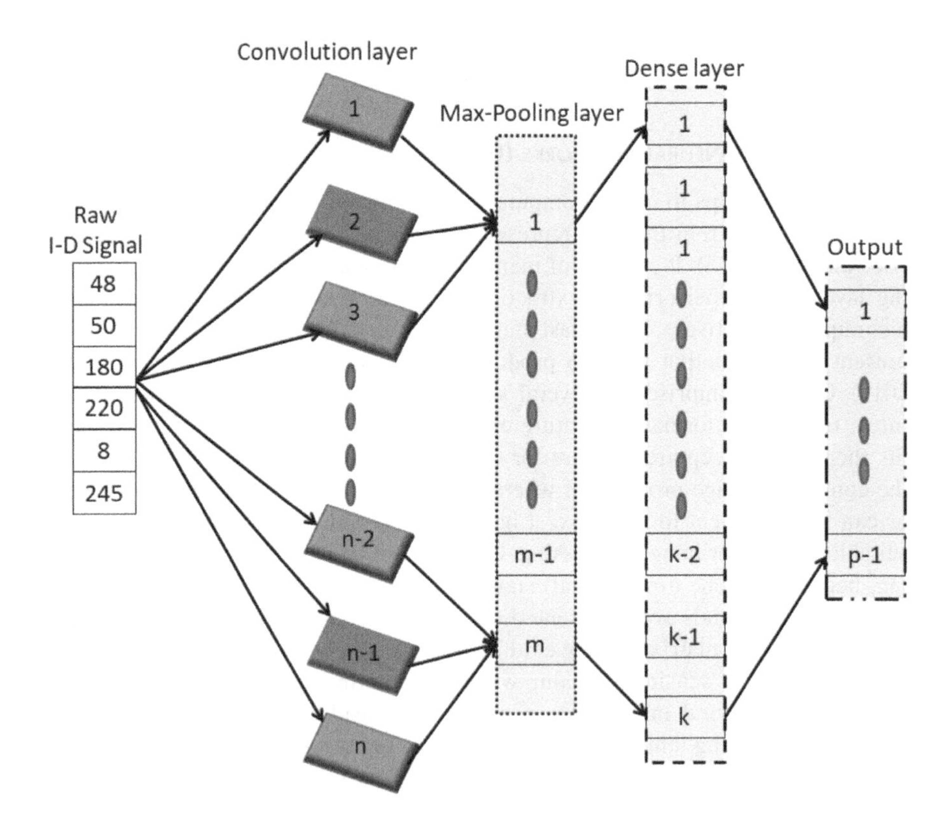

FIGURE 6.5 Structure of 1D-CNN.

Convolution Layer

The convolution layer is generally the starting layer of the network. In 1D-CNN, the convolution operation is one-dimensional to generate 1D feature maps corresponding to each convolutional operation, while feature extraction takes place through convolution kernels. The application of kernels is to detect specific features from all input features for effective weight sharing. Training-parameter size is lowered as a result of the local-connection characteristic for effective weight sharing. Because of the characteristics of local connectivity and good weight sharing, the size of training parameters is minimised, and the network becomes simpler (Huang et al., 2019). Mathematically, 1D-convolution layer of 1D-CNN is mentioned in Eq. (6.1):

$$x_j^l = f\left(\sum_{i=1}^{M} x_i^{l-1} * k_{ij}^l + b_j^l\right) \tag{6.1}$$

where '*k*' signifies convolution kernels; '*j*' is basically kernel number; '*b*' is bias; '*M*' represents the number of channels for input x_i^{l-1}; '*f*()' is the activation function with * as indicated as convolution.

Pooling Layer

Convolution layers in CNN models are suffixed with pooling sub-sampling layers, which reduces the data dimensionality. After the convolution phase, there is large number of feature maps, which makes data dimensionality very large and not suitable for calculation. So, after few convolutional operations, down-sampling, called as sub-sampling, is required. According to the literature, there are two forms of sub-sampling pooling layers: max pooling and average pooling. The maximum pixel value per window in the feature map is chosen using the max-pooling sub-sampling approach, and the average pixel value from each window is chosen using the average-pooling sub-sampling strategy.

Fully Connected Layer

The last layer of the network is called the dense layer, which is actually a fully connected layer. In this fully connected layer, all neurons in a layer are connected to the neurons of the next consecutive layer for forwarding the processed information. Sigmoid and Softmax functions, as activation functions, were used in fully connected layers for classification problems. While representing the last pooling layer as k+1 has connection to the fully connected layer, the output from the fully connected layer is mentioned in Eq. (6.2) (Huang et al., 2019):

$$h(x) = f\left(w^{k+1} x^{k+1} + b^{k+1}\right) \tag{6.2}$$

where '*w*' is weight value and '*b*' is bias in fully connected layers.

Network Training

Training of network processes starts while initially providing weight values randomly with bias parameters value to be one. It's a feed-forward network, where the inputs propagate through the number of layers, such as the convolution layer followed by pooling layers and fully connected layers, from which output values are received. Error at a particular iteration of the model is estimated while calculating the difference between expected and target values. In these steps, error value consecutively computes the errors through the fully connected pooling layer and then the convolution layer, followed by the computation of the error gradient. The error is fed to the network and weights as well as bias are again computed and updated until the minimum-permissible-error value is reached. A flowchart representing the network training process is shown in Figure 6.6.

6.3 RECURRENT NEURAL NETWORKS (RNN)

Recently, researchers developed interest in RNNs, which are mathematical models in place of statistical models (Choi and Lee, 2018). Recurrent neural-network models are similar to ANN models; however, they are mostly used for predicting operations with sequential/time-series data. Deep-learning models of the RNN type are primarily used for ordinal or time-series data issues such as neural-machine translation

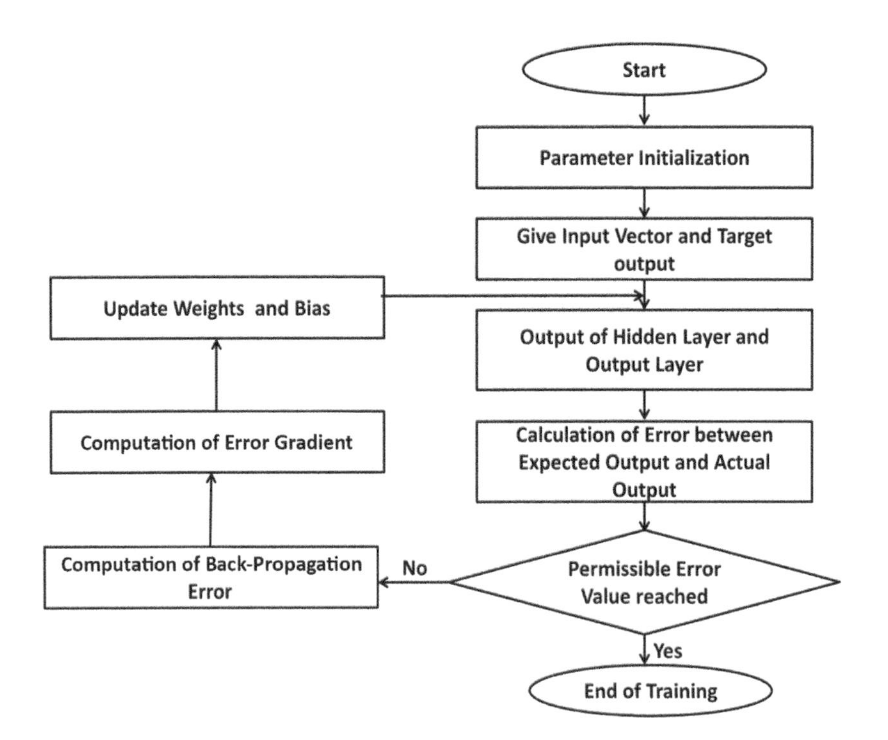

FIGURE 6.6 Network training process.

(NMT), natural-language processing (NLP), automatic image-subtitle tasks, and so on. In the present scenario, recent voice-supportive gadgets like Alexa, Google Assistance, and Siri are models for providing skills which are hassle-free to all users.

6.4 DIFFERENCE BETWEEN RNN AND CNN

Based on a comparison between CNN and RNN, in the RNN model, memory is unified so that information from previous input is considered which affects recent input, as well as output. The training approach for the RNN model is the same as for other deep-learning models. In the conventional ANN model, output is not dependent on inputs, while in RNN, output is dependent on previous inputs.

The second difference between RNN and other deep-learning models is that each layer gets shared parameters. Different weights are shared in feed-forward networks at each connectivity, while RNN shares the same weight at each connectivity with respect to each layer of the model. The structures of Feed-forward Network and Standard RNN are shown in Figure 6.7(a) and Figure 6.7(b) respectively.

With recent development in RNNs, its application has shown incredible success in various application areas, such as language modelling, image captioning, translation and speech recognition etc. Figure 6.8 shows the structure of a single cell of RNN architecture. It shows input from the previous cell, as well as from recent state X_t input, while incorporating the Tanh activation function, which can be changed also.

A number of times, only recent information has been used to execute the current task. But for sequential data, previous output has been linked with current output. So, the RNN model can link previous outputs with current output. A number of times, when RNN has been applied using long sequences or phrases, it has lost the information, as

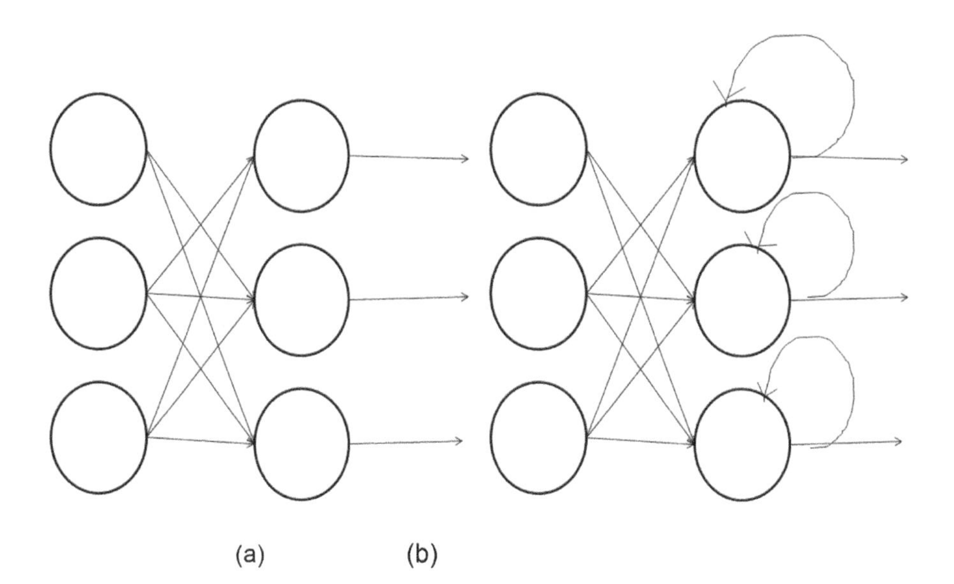

FIGURE 6.7 Structures of (a) feed-forward network (b) standard RNN.

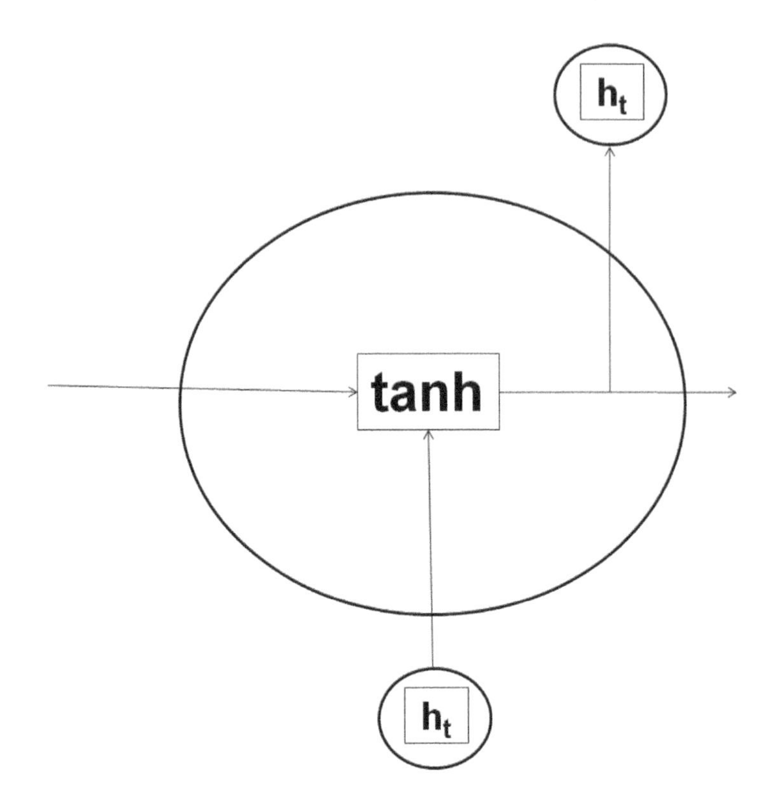

FIGURE 6.8 Single unit of RNN architecture.

RNN cannot store long sequences, but based on its characteristics, it works only using current iterative processes on a node. This problem is called vanishing gradients.

During the training of an RNN model, a time-domain back-propagation technique is used, which means that the output of iterative phases is considered as input. The loop-operation gradient is computed during each time-based iterative process, and this gradient aids in the updating of network weights. Because the effect of the previous sequence on layers is minor, the computation of relative gradient is small. As a result, because the gradient of the preceding layer is fairly minor, only a tiny amount of weight should be allocated. When longer sequences are used, the same effect is observed. This causes problems in the network by preventing it from learning from early inputs. This causes a short-term memory difficulty in the network. To triumph over this problem, very specialised types of RNN are developed, called ConvLSTM2D layer, Time Distributed layer, and GRU(Gated Recurrent Unit). Out of these four types of RNN models, LSTM (Long Short-Term Memory) and GRU (Gated Recurrent Units), are very common and will be discussed in upcoming sections.

6.5 LONG SHORT-TERM MEMORY (LSTM)

When there is the issue of learning long-term sequences, also called long-term data-dependency in RNN, then LSTM is a special type of RNN, proficient in learning long-term sequences (Figure 6.9). This happens due to LSTM increases memory in the RNN model. Hochreiter and Schmidhuber introduced LSTM in 1997. It was specially designed to handle the long-term dependency issues. The advantage of LSTM is memorising long sequences for a long duration.

Due to the involvement of mechanisms in each cell of LSTM, it has become popular in short span of time. In an ordinary RNN unit, the input from current iteration and output from previous iteration is passed to an activation layer for obtaining a new form, while, in LSTM, the same process is slightly more complicated, as shown in the Figure 6.9. At every iteration, LSTM receives input from three stages, which are the present-input stage, from previous cell short, as well as the last long-term memories.

The concept of passing and regulating information by cells is through the gate. So that the information is considered or discarded during the loop operation, before categorized as short- or long-term and passed to subsequent cells. The role of gates is like strainers, which remove unwanted selected/irrelevant information. Three gates are there in LSTM, which are: Input, Forget, and Output Gate.

Input Gate

Based on the working of input gate, it is decided which information is to be stored as long-term memory. Input-gate use basically presents iteration input, as well as short-term memory generated from the preceding iteration. Input gate filters out non-useful information from variables.

Forget Gate

Forget gate decides whether to accept or discard information from long term memory. There is a forget vector which makes it possible to achieve this information using

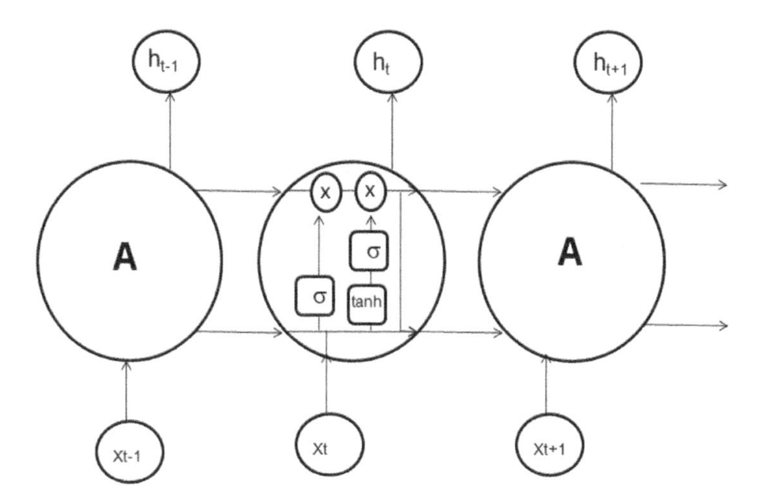

FIGURE 6.9 Long Short-Term Memory (LSTM) structure (https://colah.github.io/posts/2015-08-Understanding-LSTMs/).

current input with incoming short-term memory, which is multiplied with long-term memory.

Output Gate

The role of output gate is to generate new short-term memory, using present input and earlier short-term memory with newly calculated long-term memory. The generated, new short-term memory is further used by cells present in the next iteration, while output of the current iteration is drawn through a hidden state.

Long Short-Term Memory (LSTM) form of RNN is a very different type. LSTM has been applied in different application areas, like machine translation, speech recognition, text recognition, etc. (Ju et al., 2017). LSTM handles the vanishing-gradient problem (Hochreiter and Schmidhuber, 1997). LSTM was further expanded by Gers in 2001. Sequence data of a large time-period having context information can be easily memorised by the LSTM model (Byeon et al., 2015).

Some of the parameters in LSTM can be defined as input vector with $'X_t'$; output from previous cell as $'M_{t-1}'$; previous cell memory value as $'N_{t-1}'$; current cell output as $'M_t'$; current cell memory as $'N_t'$. From these LSTM parameters, LSTM cell equations have been represented by Eq. (6.3), (6.4), (6.5), (6.6), (6.7), and (6.8) (Hochreiter and Schmidhuber, 1997; Tai et al., 2015):

$$f_t = \tilde{A}\left(X_t * U_f + M_{t-1} * W_f\right) \tag{6.3}$$

$$\bar{N}_t = tanh\left(X_t * U_c + M_{t-1} * W_c\right) \tag{6.4}$$

$$I_t = \sigma\left(X_t * U_i + M_{t-1} * W_i\right) \tag{6.5}$$

$$O_t = \sigma\left(X_t * U_o + M_{t-1} * W_o\right) \tag{6.6}$$

$$N_t = f_t * N_{t-1} + I_t * \bar{N}_t \tag{6.7}$$

$$M_t = O_t * \tanh(N_t) \tag{6.8}$$

In these equations, forget gate is represented as '*f*'; this is equivalent to sigmoid activation-function-based neural network; 'tanh' is the activation function in the neural network; '*C*' is candidate date; input gate is denoted as '*I*' (neural network with sigmoid activation function); output gate, as '*O*' (sigmoid activation function in neural network); hidden state as a vector '*M*'; memory state is considered as a vector '*N*'; time step is 'T'; for all gates, weight vectors are depicted as 'U' and 'W' (Figure 6.10).

6.6 GATED RECURRENT UNIT (GRU)

Gated Recurrent Unit (GRU) follows the same steps as the RNN model, however it differs in process and types of gates linked with every GRU cell (Figure 6.11). In a standard RNN, there were few problems while in a GRU, gates are incorporated called Reset gate and Update gate.

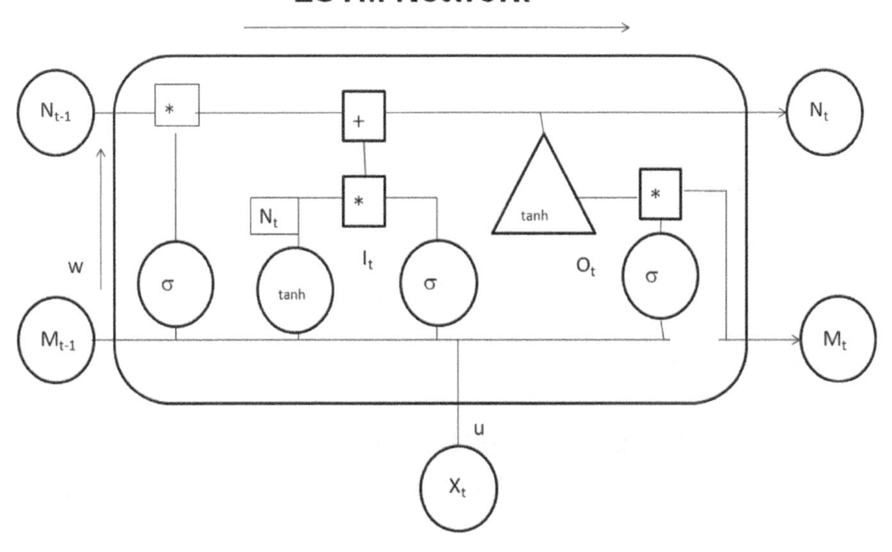

FIGURE 6.10 An LSTM cell.

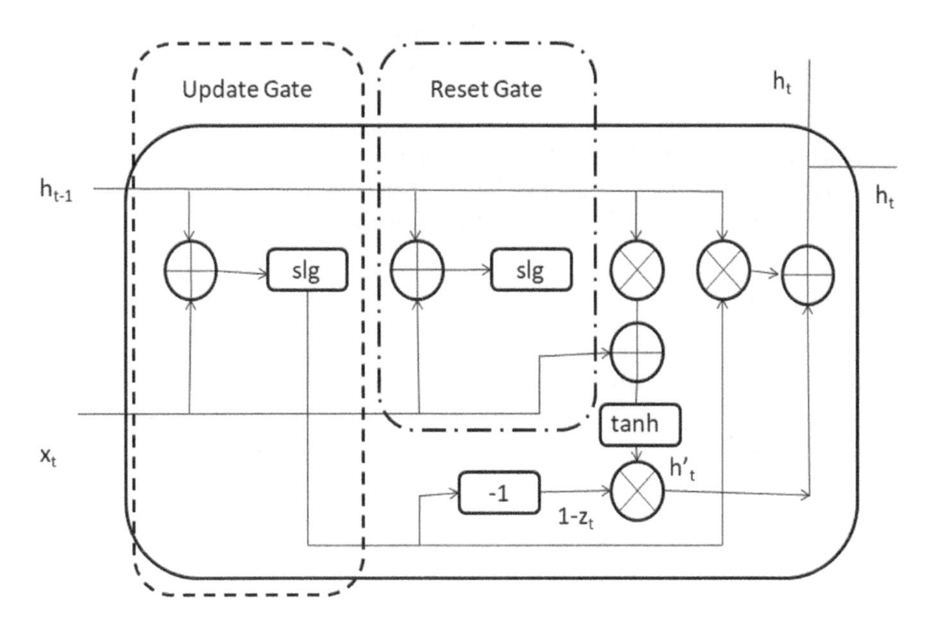

FIGURE 6.11 A Gated Recurrent Unit (GRU).

Update Gate

The role of update gate is to find out past iteration-information that has to be added with next iteration-information. This gives power to the model to decide and copy previous iteration information to remove the jeopardy related to the vanishing gradient.

Reset Gate

The use of reset gate in the GRU model is to find out, how much information of past iteration has to be neglected. On the other hand, reset gate decides whether a previous-iteration cell-state is important or not.

The main steps followed in reset gate are: to store pertinent information from a past iteration into new memory; in the next step, a hidden state is multiplied with the input vector while considering their corresponding weight values; later, component-wise multiplication takes place among previously hidden-state multiples and reset gates. All these steps are summed up and passed into the activation function as next-sequence outputs.

6.7 DIFFERENCE BETWEEN GRU & LSTM

Some of the differences between GRU and LSTM can be listed here, such as: in GRU, there are two gates, while there are three gates in LSTM; there is no internal memory in GRU; this is due to no output gate being present, however in LSTM, an output gate is present.

In the case of LSTM, update gate is used to fix input gate, as well as target gate. While, in the case of GRU, previous hidden-state reset gates are directly applied. Two input and target gates are responsible for reset gate in LSTM.

While working with LSTM and GRU layers, it has been found that GRU takes fewer training parameters and hence usage less memory. GRU also executes faster in comparison to LSTM. LSTM is more accurate while working on a larger-size or large sequences of data. On the other hand, when there is less memory and results need to be achieved faster, the GRU model can be used.

6.8 SUMMARY

In this chapter, the background of deep learning has been presented with a thorough presentation of the workings of the CNN model, with a focus on the 1D-CNN sequential model and a comparison with the RNN model. Various variants of the RNN model have also been discussed.

BIBLIOGRAPHY

Byeon, W., Breuel, T. M., Raue, F., & Liwicki, M. (2015). Scene Labeling with LSTM Recurrent Neural Networks. *IEEE Conference on Computer Vision and Pattern Recognition (CVPR)*, 3547–3555. Boston.

Choi, J., & Lee, B. (2018). Combining LSTM Network Ensemble via Adaptive Weighting for Improved Time Series Forecasting. *Mathematical Problems in Engineering*, *2018*, 1–8. https://doi.org/10.1155/2018/2470171.

Fogg. (2019). *A History of Machine Learning and Deep Learning | Import.io*. Retrieved December 20, 2019, from www.import.io/post/history-of-deep-learning/

Hochreiter, S., & Schmidhuber, J. (1997). Long Short-term Memory. *Neural Computation*, *9*, 1735–1780. https://doi.org/10.1162/neco.1997.9.8.1735.

Huang, S., Tang, J., Dai, J., & Wang, Y. (2019). Signal Status Recognition Based on 1DCNN and Its Feature Extraction Mechanism Analysis. *Sensors*, 1–19. https://doi.org/10.3390/s19092018.

Ju, C., Bibaut, A., & Laan, M. J. Van Der. (2017). The Relative Performance of Ensemble Methods with Deep Convolutional Neural Networks for Image Classification. *Journal of Applied Statistics*, *45*(15), 1–20. https://doi.org/10.1080/02664763.2018.1441383.

Kiranyaz, S., Avci, O., Abdeljaber, O., Ince, T., Gabbouj, M., & Inman, D. J. (2019). 1D Convolutional Neural Networks and Applications: A Survey. *Mechanical Systems and Signal Processing*, *151*, 1–20. https://doi.org/1905.03554v1.

LeCun, Y., Bengio, Y., & Hinton, G. (2015). Deep Learning. *Nature*, *521*(7553), 436–444. https://doi.org/10.1038/nature14539.

Liang, H., & Li, Q. (2016). Hyperspectral Imagery Classification using Sparse Representations of Convolutional Neural Network Features. *Remote Sensing*, *8*(2). https://doi.org/10.3390/rs8020099.

Maggiori, E., Tarabalka, Y., Charpiat, G., & Alliez, P. (2017). Convolutional Neural Networks for Large-Scale Remote-Sensing Image Classification. *IEEE Transactions on Geoscience and Remote Sensing*, *55*(2), 645–657. https://doi.org/10.1109/TGRS.2016.2612821.

Michalski, R., Carbonell, J., & Mitchell, T. (1983). *Machine Learning. An Artificial Intelligence Approach*. Volume 2. Springer-Verlag, Berlin Heidelberg GmbH. https://doi.org/10.1007/978-3-662-12405-5.

Ndikumana, E., Minh, D. H., Baghdadi, N., Courault, D., & Hossard, L. (2018). Deep Recurrent Neural Network for Agricultural Classification Using Multitemporal SAR Sentinel-1 for Camargue, France. *Remote Sensing*, *10*(8), 1–16. https://doi.org/10.3390/rs10081217.

Pelletier, C., Webb, G. I., & Petitjean, F. (2019). Temporal Convolutional Neural Network for the Classification of Satellite Image Time Series. *Remote Sensing*, *11*(5), 1–25. https://doi.org/10.3390/rs11050523.

Rubwurm, M., & Korner, M. (2017). Temporal Vegetation Modelling Using Long Short-Term Memory Networks for Crop Identification from Medium-Resolution Multi-spectral Satellite Images. *IEEE Computer Society Conference on Computer Vision and Pattern Recognition Workshops*, *2017-July*, 1496–1504. Publisher IEEE. https://www.semanticscholar.org/paper/Temporal-Vegetation-Modelling-Using-Long-Short-Term-Ru%C3%9Fwurm-K%C3%B6rner/f5652d500676829ec997f7dd27dea672f458a4d1.

Saha, S. (2018). *A Comprehensive Guide to Convolutional Neural Networks—the ELI5 Way*. Retrieved December 20, 2019, from https://towardsdatascience.com/a-comprehensive-guide-to-convolutional-neural-networks-the-eli5-way-3bd2b1164a53.

Tai, K. S., Socher, R., & Manning, C. D. (2015). Improved Semantic Representations from Tree-Structured Long Short-Term Memory Networks. *ACL-IJCNLP 2015–53rd Annual Meeting of the Association for Computational Linguistics and the 7th International Joint Conference on Natural Language Processing of the Asian Federation of Natural Language Processing, Proceedings of the Conference*, *1*, 1556–1566.

Vargas, R., Mosavi, A., & Ruiz, R. (2018). Deep Learning: A Review. *Advances in Intelligent Systems and Computing*. https://www.preprints.org/manuscript/201810.0218/v1.

Yoo, H. (2015). Deep Convolution Neural Networks in Computer Vision: A Review. *IEIE Transactions on Smart Processing and Computing*, *4*(1), 35–43. https://doi.org/10.5573/IEIESPC.2015.4.1.035.

Zhang, G., Xiao, X., Dong, J., Kou, W., Jin, C., Qin, Y., & Biradar, C. (2015). Mapping Paddy Rice Planting Areas Through Time Series Analysis of MODIS Land Surface Temperature and Vegetation Index Data. *ISPRS Journal of Photogrammetry and Remote Sensing*, *106*, 157–171. https://doi.org/110.1016/j.isprsjprs.2015.05.011.

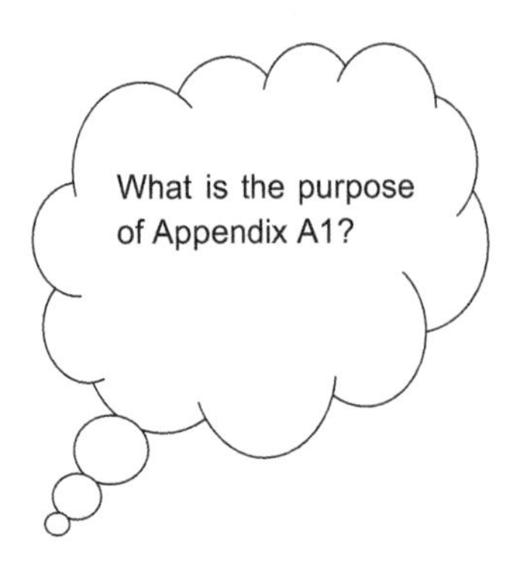

To provide insight into specific single-class mapping from multi-temporal and single/dual/multi-sensor remote sensing images.

. . . .Begin your day with. . . .
Gratitude. . . .
With Positive Vision. . . .
Have Broader Plans. . . .
Do Small Things Right Way. . . .
Help Someone. . . .

Appendix A1
Specific Single Class Mapping Case Studies

Various research works describing specific single class mapping while involving dual or multi-sensor temporal data with fuzzy or deep-learning models have been described and discussed in this section.

A1. FUZZY VERSUS DEEP-LEARNING CLASSIFIERS FOR TRANSPLANTED PADDY FIELDS MAPPING

To achieve the aims of this research, a study region featuring paddy fields was identified around nearby cities/towns, such as Ambala, Pehowa, Kurukshetra, Ladwa, Radaur in the state of Haryana and Punjab, India (Figure A1.1).

The study area's boundary coordinates are 30°26′N, 77°11′E top right, with 29°56′N, 77°10′E bottom left. In this study area, the main crop in the Kharif season is paddy. Maize and sugarcane are two other crops grown along with paddy in this region.

The temporal-data technique was used to incorporate phenological stages of a rice crop in this research case study. In order to fill the temporal gaps, multi-sensor data have been used. Mainly freely available data of Sentinel-2 were used due to its high revisit cycle. Landsat-8 satellite data was used to reduce temporal spacing within Sentinel-2 temporal data. As paddy is grown in the monsoon season, so availability of optical data is the main concern. Further microwave data from the Sentinel-1 has also been incorporated. The sensor details and specifications about temporal data used have been listed in coming sections.

Sentinel-2 Temporal Data Details

Sentinel-2 which is under the Copernicus program a European Space Agency (ESA) mission intended for Earth observation. Being an optical remote-sensing sensor, it acquires images to be of medium-finer spatial resolution. The sentinel satellite mission is a dual-satellite constellation, having satellites Sentinel-2A/2B. The Sentinel-2A/2B image swath is 290 km, and due to their being a dual-constellation satellite, their revisit time is five days. This temporal-revisit time is due to two identical satellites called Sentinel-2A/-2B. The radiometric resolution of Sentinel-2 image is 12 bits. Copernicus Open Access Hub Sentinel-2A/2B data can be explored and be downloaded freely. Atmospherically corrected level-2 images from Sentinel-2A/2B do not require further pre-processing steps. A specification chart depicting Sentinel-2 bands and their spatial resolution is shown in Table A1.1.

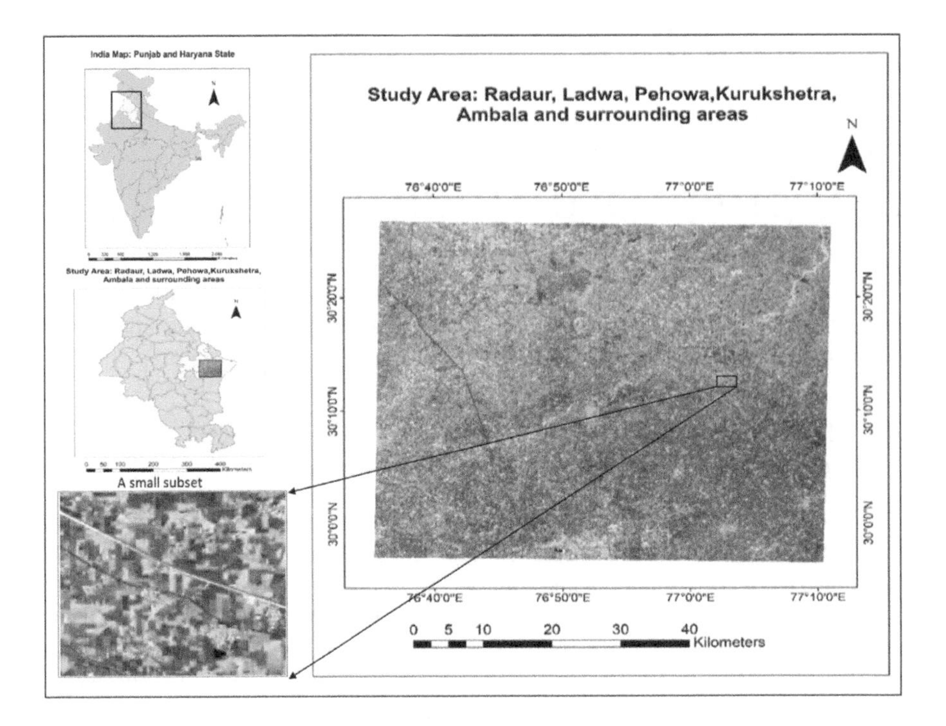

FIGURE A1.1 Geographic location of the study area.

TABLE A1.1
Band combinations of Sentinel-2A/2B.

Sentinel-2 Bands	Central Wavelength (μm)	Resolution (m)
Band 1-Coastal aerosol	0.443	60
Band 2-Blue	0.490	10
Band 3-Green	0.560	10
Band 4-Red	0.665	10
Band 5-Vegetation Red Edge	0.705	20
Band 6-Vegetation Red Edge	0.740	20
Band 7-Vegetation Red Edge	0.783	20
Band 8-NIR	0.842	10
Band 8A-Vegetation Red Edge	0.865	20
Band 9-Water vapour	0.945	60
Band 10-SWIR-Cirrus	1.375	60
Band 11-SWIR	1.610	20
Band 12-SWIR	2.190	20

TABLE A1.2
Band combinations of Landsat-8.

Sentinel-2 Bands	Wavelength Range (μm)	Resolution (m)
Band 1-Coastal/aerosol	0.433–0.453	30
Band 2-Blue	0.450–0.515	30
Band 3-Green	0.525–0.600	30
Band 4-Red	0.630–0.680	30
Band 5-Near Infrared	0.845–0.885	30
Band 6-Short Wave Infrared	1.560–1.660	30
Band 7-Short Wave Infrared	2.100–2.300	30
Band 8-Panchromatic	0.500–0.680	15
Band 9-Cirrus	1.36–0–1.390	30

Landsat-8 Temporal Data Details

It was launched by Orbital Science Corporation. Landsat-8 satellite sensor is a sun-synchronous, near-polar satellite having a revisit cycle of 16 days. The scene size is 170km ×185km. The radiometric resolution of Landsat-8 is 12 bits. Its data is freely available at USGS through Earth Explorer. It consists of nine spectral bands. For Landsat-8, level-2 atmospherically corrected data was also acquired. A detailed specification chart depicting Landsat-8 spectral bands along with their spatial resolution has been mentioned in Table A1.2.

Sentinel-1 Data Details

Sentinel-1 is the first satellite of the Copernicus program maintained by the European Space Agency. Like Sentinel-2, it also consists of twin satellites, Sentinel-1A and 1B. They have a payload of C-Band SAR (Synthetic Aperture Radar). The spatial resolution varies depending on the mode. For strip-map mode, the swath is of 80 km and 5×5m spatial resolution. Sentinel-1A/1B complete 175 orbits in a cycle, with a 12-day repeat-cycle orbit. It has dual polarisation mode VV+VH, HH+HV. Here, the first H means 'horizontally transmitted' and the second H means 'horizontally received'. First V means 'vertically transmitted' and the second V means 'vertically received'. It gives two products: the GRD (Ground Range Detected) and SLC (Single Look Complex). SLC data gives information with phase and amplitude as complex images. GRD data gives systematically distributed, multi-looked intensity.

In this case study, multi-sensor temporal data was used, including Sentinel-1, 2 and Landsat-8 data sets. Along with Sentinel-2 optical data, Landsat-8 optical data and Sentinel-1 microwave data were also used. The available multi-sensor temporal data used for this study has been listed in Table A1.3.

A field trip to the research area was made to collect training and testing ground-truth data on 27th and 28th July 2019 (Figure A1.2). It was conducted in selected

TABLE A1.3
Available temporal data.

Temporal Data	Sensors
6th March; 5th May; 25th May; 30th May; 9th June; 14th June; 19th June; 2th June 2019	Sentinel-2 (10m)
8th July 2019	Landsat-8 (30m)
13th July 2019	Sentinel-GRDH(10×10m)

regions of Ambala, Radaur, Kurukshetra, Pehowa, and Ladwa towns of the Haryana and Punjab state. Geo-location and the information of various LULC (Landuse/Landcover) classes for the study area was collected. During a field visit, paddy fields of large size were identified and farmers were consulted to find out paddy transplantation dates.

Human experience, as one type of cognitive-science approach, was used on known, forthcoming rice-crop training fields. For example, training fields identified on 8th July 2019 were considered as information based on experience to identify the transplanted paddy fields' dates in the last few years. Because of the higher moisture content in the typical False Colour Composite (FCC) image, transplanted paddy fields appear dark green (blackish) in the time domain. The same paddy fallow lands that will be transplanted in the near future appear light green, since they were dry, while transplanted paddy fields appear pinkish, owing to biomass increase. With these changes in rice fields over time, pinkish paddy

FIGURE A1.2 Paddy transplanted on 27th July; paddy transplanted 15 days ago; paddy transplanted one month ago.

fields on 8th July appear to have been transplanted in the previous few weeks. As a result, previously transplanted paddy fields with sufficient grown-up biomass emerge as dark green on 8th July 2019, while the same paddy field before transplantation appear to be very light green on standard FCC. Similarly, examining paddy fields that look reddish on normal FCC dates reveals that paddy was transplanted earlier in these farms. Overall, the information gathered during field observations was utilised as a cognitive-science technique. In the past, this cognitive science was employed to identify training and evaluating transplanted paddy-field samples. Through this approach, training and testing samples for varied periods of paddy field transplantation have been obtained for the complete paddy transplantation times in a session. The geolocations of these fields were also captured.

Figure A1.3 illustrates the flow diagram of the approach used to conduct this investigation. Following section provides a full explanation of the methodology as well as all of the pre-processing, processing, and post-processing procedures used to carry out this study.

One of the primary goals of this research was to determine the role of indices in temporal-data processing and to identify the best temporal-date combination for mapping a specific crop. To do so the images from all sensors were pre-processed in accordance with the standards of optical and microwave data pre-processing. Then, using all the temporal images available the temporal indices database was generated. Following that, a separability study was undertaken to select the optimum-date

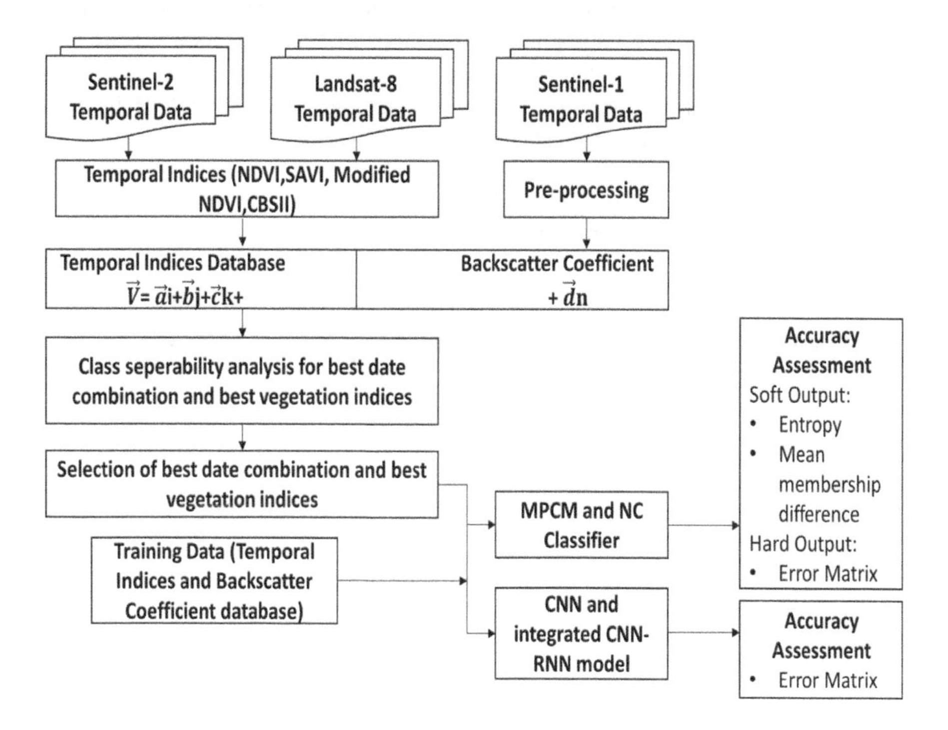

FIGURE A1.3 Methodology adopted for this case study.

combination from among two, three, and four date combinations. This helped to conclude the most appropriate vegetation index and the best date combination which is able to provide the maximum separation among the classes. Then the best vegetation-index images for the best date combination were stacked together and training sites were confirmed to train the classifier. The same training sites were used to train all the classifiers, whether they are fuzzy-based or learning-based; only difference is the number of training samples as per the requirement of classifier. For each classifier, the optimum value of parameters were determined. Thereafter, the paddy crop was classified according to the date of transplantation. Hard outputs were classified and evaluated using the error-matrix approaches in relation to learning algorithms. Along with the error-matrix computation on the hard output the mean membership difference and entropy were produced as an accuracy metric for the fuzzy-based approach. This was done in order to determine the optimal categorisation strategy for specific-crop mapping using multi-temporal investigations.

MPCM and NC fuzzy classifiers implemented in this research case study can be referred to from Chapter 5.

In this case study, the 1D-CNN sequential model, an integrated 1D-CNN-RNN model consisting 1D convolution layers, and LSTM layers were tested. The CNN and LSTM layers in the deep-learning model were included as three convolution layers and three LSTM layers. Neurons of size 32, 64, and 128 in CNN layers and 64, 128, and 256, in LSTM layers were used. In convolution layers, the Relu activation function was used. Dropout rates of 0.6, 0.5, and 0.4 were considered in three dropout layers, respectively. Adam optimiser was used, having a 0.001optimum learning rate in CNN and 1D-CNN-LSTM models. Softmax activation function was used in the dense layer. Optimised batch sizes, as well as an optimised number of epochs considered, were 10 and 150, respectively. CNN and integrated 1D-CNN-RNN optimised-model structures have been shown in Figure A1.4 and A1.5.

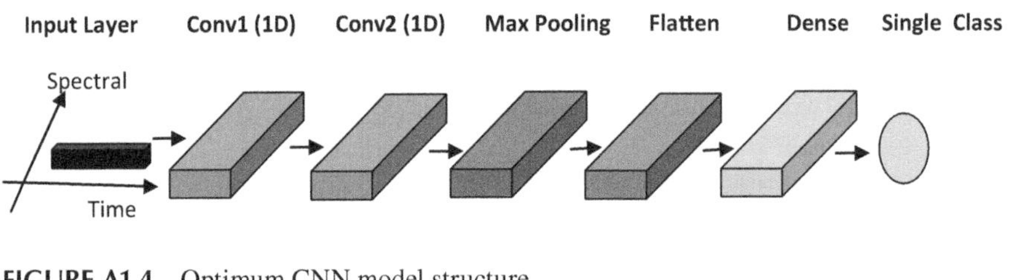

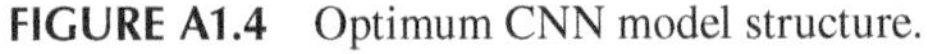

FIGURE A1.4 Optimum CNN model structure.

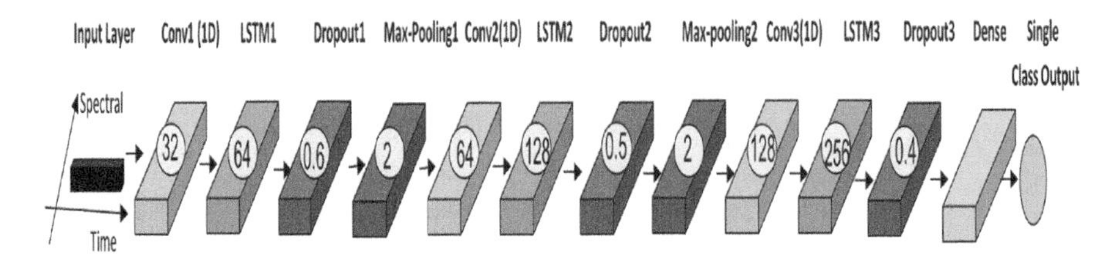

FIGURE A1.5 Optimum, integrated CNN-RNN model structure.

In this research study, transplanted paddy fields have been mapped on 9th July 2019, while using multi-sensor images. While using two dates' data of Sentinel-2, there was one image from Landsat-8 on 8th July 2019, during which paddy transplantation was mapped. Sentinel-1 image dated 13th July 2019 representing post-transplantation stage of paddy fields. But, with these four temporal-date data, a non-perennial canal mapped as a transplanted paddy field (Figure A1.6). To handle this false non-perennial canal, one more image has been added of 6th March 2019, representing a wheat crop in the majority of this temporal data, which made possible to map only transplanted paddy fields, as shown in Figure A1.7.

The first goal of this research was to demonstrate that the CBSI indices approach produced the best separability findings of transplanted paddy fields while employing temporal-domain information. The CBSI technique analyses bands in such a way that the indices value for the class of interest is maximised in every temporal image. In this research study, while investigating fuzzy NC and MPCM models, it was discovered that both these models performed equally well. When comparing CNN and CNN-LSTM deep-learning models, it was found that the CNN model outperformed. Furthermore, the CNN model necessitates fewer model layers and less hyper-parameter adjustment. Out of the four models tested, the CNN also outperformed the other models. It was also discovered that 50 training samples were sufficient for training the CNN model and achieving the best results.

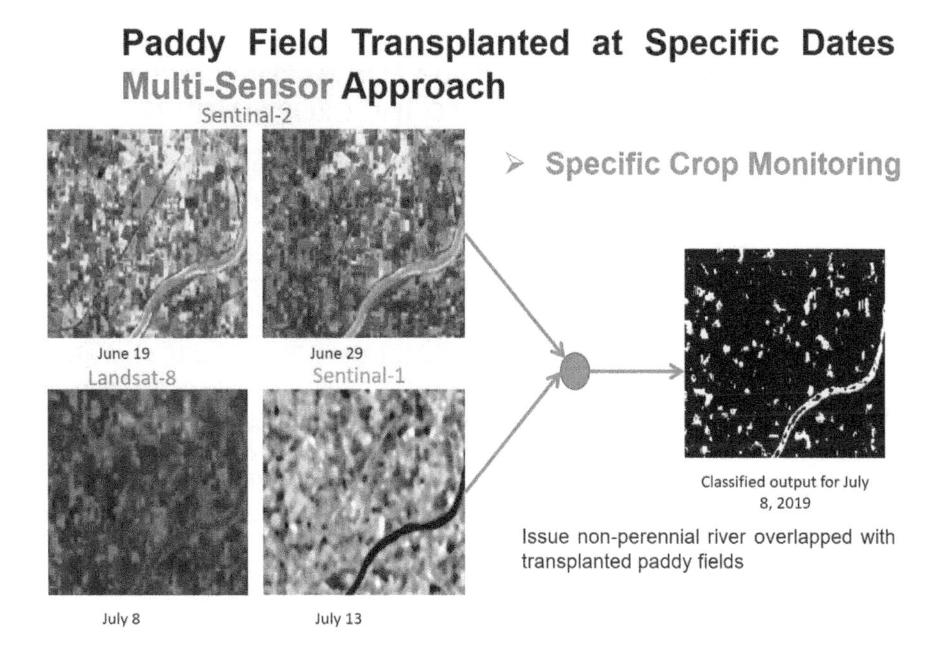

FIGURE A1.6 Transplanted paddy-fields mapping from multi-sensor and multi-temporal data.

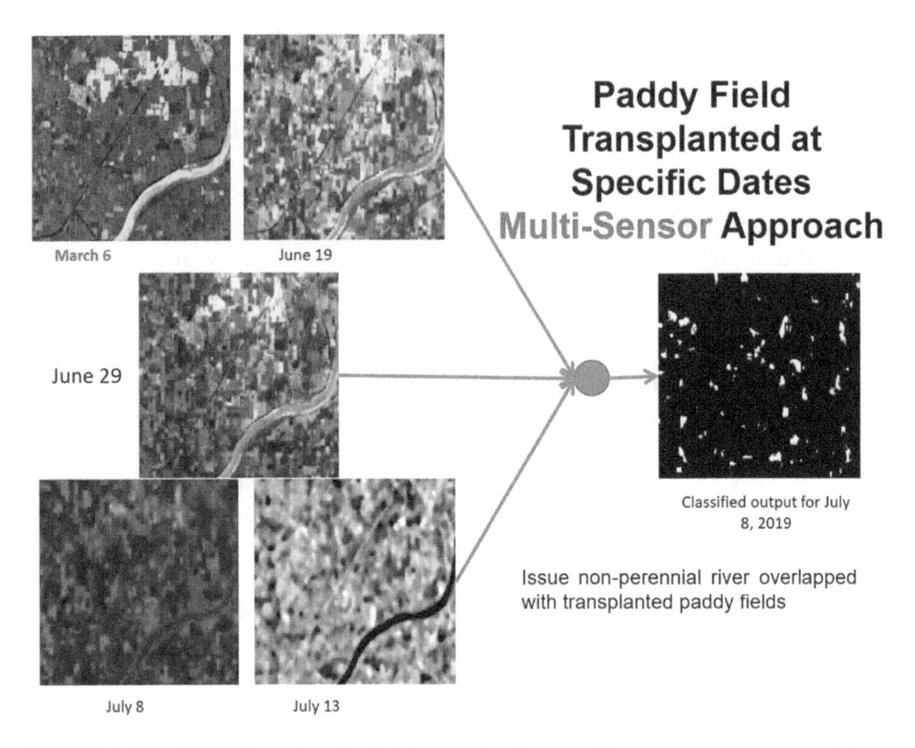

FIGURE A1.7 Transplanted paddy-fields mapping from multi-sensor, multi-temporal data, while removing dry canal from the study area.

A2. DUAL-SENSOR TEMPORAL DATA FOR MAPPING FOREST VEGETATION SPECIES AND SPECIFIC-CROP MAPPING

Dual-sensor (optical and SAR) temporal data were used in this research study for customised crop and vegetation mapping. In this study two test sites one is in Sri Lanka and other is in India were selected for this purpose. Sentinel-2A/2B satellite temporal-data covers many bands with varying spatial resolutions. For subsequent processing, all of the multi-spectral bands were resampled to a pixel size of 10m. The re-sampled data set was utilised to generate the temporal-indices database.

Radiometrically and atmospherically corrected ortho-rectified data (L2A product) were resampled. This was done for all 20m spatial-resolution bands of the temporal images at both study areas. The MSI sensor of the Sentinel-2A/2B satellite have blue, green, red and near-infrared bands with 10m resolution and three red-edge bands with 20m resolution. All the red-edge bands were re-sampled to 10m using the nearest-neighbour re-sampling method. A new layer-stacked image was created with all the concerned bands of 10m resolution for each temporal date.

Sentinel 1A and 1B microwave images of Interferometric Wide Swath (IWS)—Ground Range Detected (GRD) level 01 images—were used for pre-processing. Orbit correction was done for the Sentinel-1 1W-GRD data. Orbit-correction images were taken for the radiometric correction. Output-calibrated products were saved in dB-scale. Speckle filtering was applied by using Boxcar (mean) filter for the images.

Later, geometric correction was done by using SRTM DEM. Output products were converted to linear scale to decibel.

For Sentinel 2A and 2B, Class-Based Sensor Independent (CBSI)—Normalised Different Vegetation Index (NDVI)—technique was adopted to generate indices as in Eq. (7.1). Layer-stack optical images were used for both sites to find out the best dates for database generation. Separability analysis was done to sort out the best temporal CBSI-NDVI images. It aided in the differentiation of various vegetation types and crop verities in Sri Lankan and Indian sites respectively. Selected temporal images were transformed to binary format as output.

$$CBSI - NDVI = \frac{max - min}{max + min} \qquad \text{Eq. (7.1)}$$

In Eq. 7.1, '*max*' and '*min*' represent bands having 'maximum' and 'minimum' reflectance values at a given interested class location. Band values received as 'maximum' represent the NIR band, and those received as 'minimum' value represent the red band.

The temporal data containing agricultural-indices information can illustrate spectral variations in a crop over time. Knowledge about the sorts of bands contained in multi-spectral images should be known in order to generate indices databases. Obtaining band information for the purpose of generating certain indices necessitates specialist

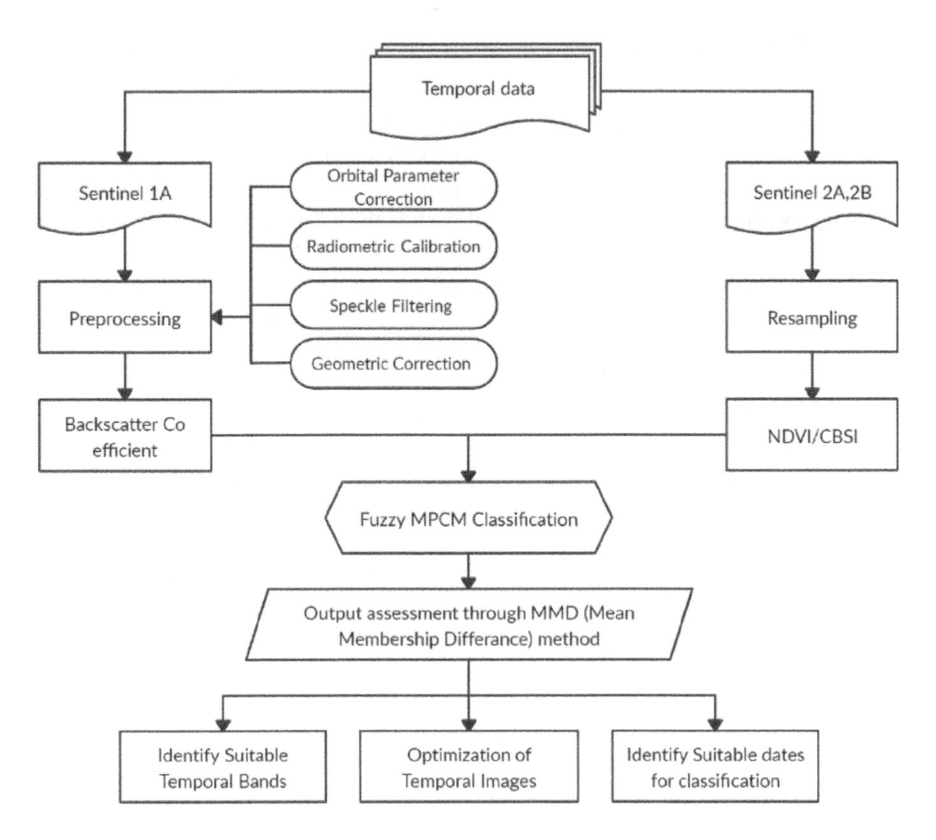

FIGURE A1.8 Proposed methodology of case study A2.

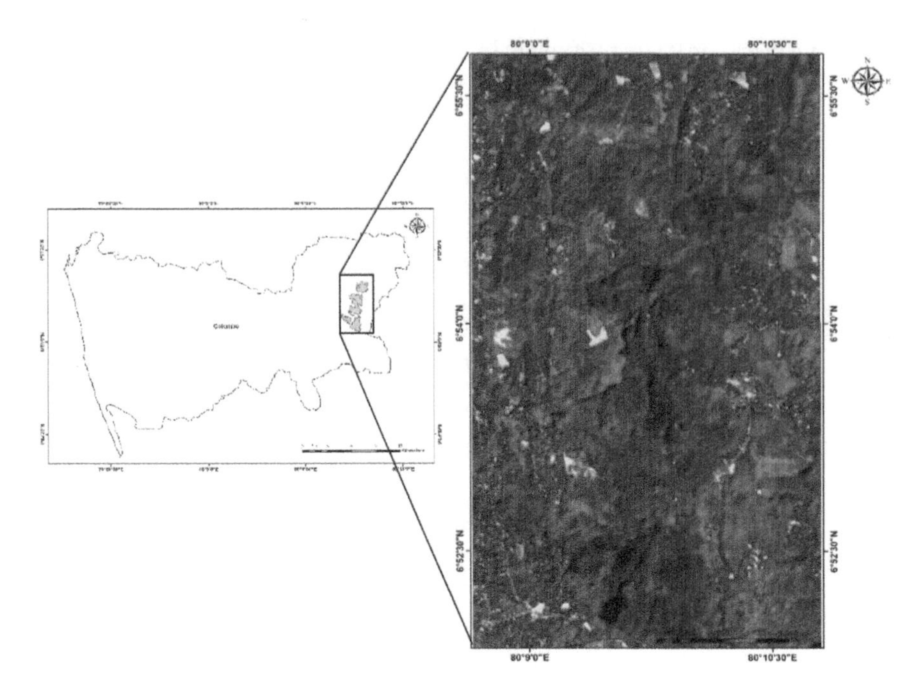

FIGURE A1.9 Indikada-Mukalana Forest (IMF) at Waga, Colombo district, Sri Lanka.

knowledge of the bands present in a multi-spectral image. To avoid the need for knowledge of the sorts of bands present, the CBSI approach supplies acceptable bands necessary in an indices formula, ensuring that the class of interest receives maximum enhancement in any temporal image (Upadhyay et al., 2012). Only the location of a class in the form of a row and column or latitude/longitude is required for CBSI-NDVI creation.

The supervised MPCM classification approach was used in this investigation. The classifier selected for the study is Modified Possibilistic *c*-Means (MPCM) (Figure A1.8). CBSI-NDVI images of different classes were used for MPCM classification. Optical output of CBSI-NDVI images were combined with the nearest date of the SAR images, while adding one SAR image at a time. Likewise, three SAR images were added with the CBSI-NDVI images.

The first study site comprised the Indikada-Mukalana Forest (IMF) at Waga, in the Colombo district of the Western Province of Sri Lanka (Figure A1.9) (Dela and Padmalal, 2018). This forest spans about 573 hectares (overall) and is now a gazetted Forest Reserve, although it was logged in the 1970s (Dela and Padmalal, 2018). It consists of four separate blocks, surrounded by rubber plantations, a 50ha coconut plantation, and village home-gardens. Two forest blocks have natural, secondary, lowland rainforest in various stages of regeneration and a small remnant-patch of remnant Pinus plantations (Dela and Padmalal, 2018). Two blocks are termed forest plantations, as they have patches of monoculture with *Dipterocarpus zeylanicus* (hora) and mixed stands of Pinus and Araucaria. The natural forest in Block I was highly degraded and some areas were heavily invaded with 'para' *Dillenia suffruticosa* (ibid). A comprehensive vegetation survey carried out by the ongoing project for conservation of the Western Purple-Faced

Langur showed that the natural forest varied significantly in terms of species due to differences in regeneration (Dela and Padmalal, 2018).

The second testing area considered was the surrounding Newai Tonk district of Rajasthan state, India. Towards its north is Jaipur district, Sawai Madhopur, and Kota districts are on the east side, Bundi district is in the south, and Ajmer district is in the west of the study area. Jaipur, the capital of Rajasthan, is the closest major city to Newai, where the second study area is located near to Vanasthali University (Figure A1.10). The Indian farming season is divided into two seasons: Rabi and

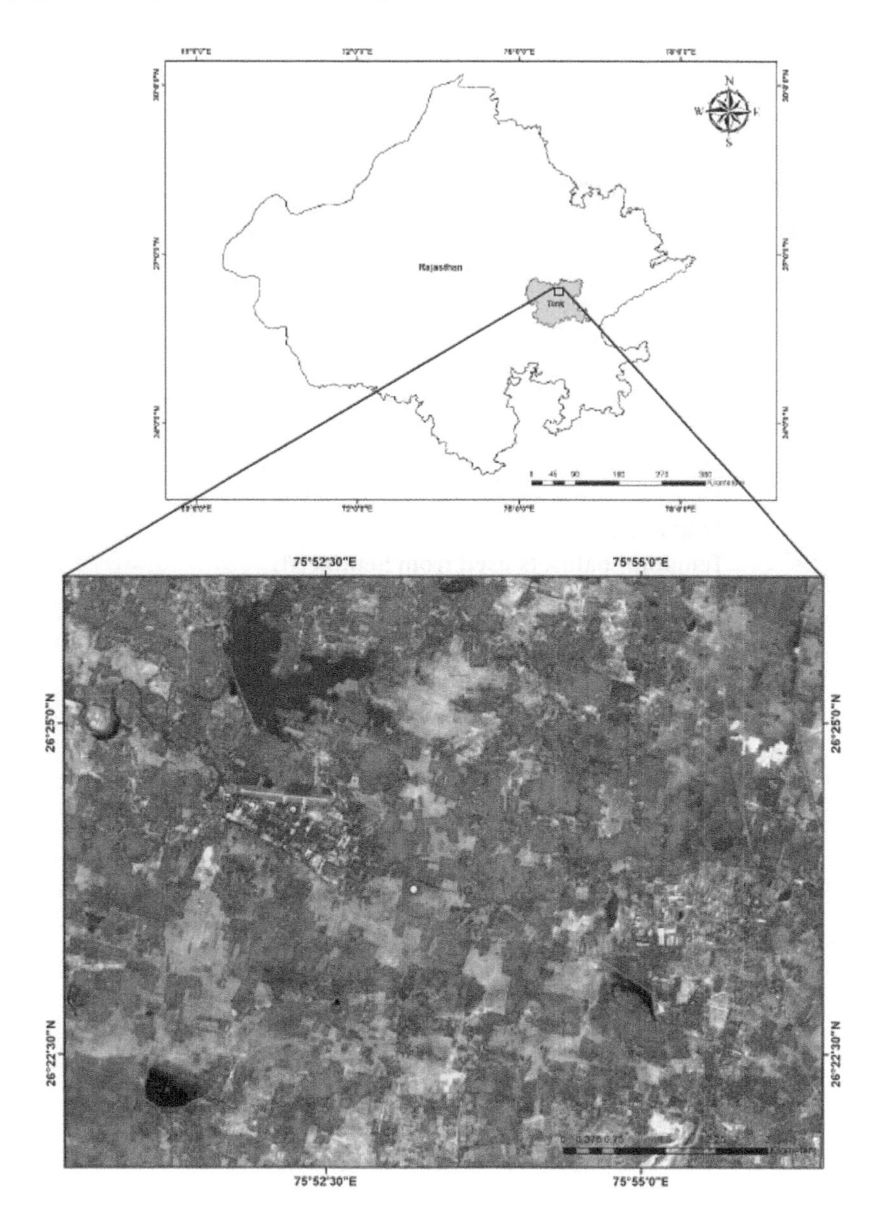

FIGURE A1.10 Vanasthali India, Study area.

Kharif. Mustard, sorghum, Kharif pulses, sesamum, and wheat are the key crops in this area that contribute to agriculture. Mustard and wheat are two important Rabi crops in this study area.

Optical remote-sensing temporal images from multi-spectral imager (MSI) of Sentinel-2A/2B satellites were acquired. For the dataset from January 2019 to January 2020, cloud-free images were used for the Sri Lankan study site. Cloud-free images from October 2019 to January 2020 were used for the Indian study site. The temporal data mentioned in Tables A1.4 and A1.5 were considered in this study. The band details of the MSI sensor have been given in Table A1.6.

TABLE A1.4
Temporal datasets used from Sentinel-2.

Study Site	Sentinel 2A/2B (L2A product)
Sri Lanka	31–01–2019, 06–01–2019, 27–11–2019, 27–03–2019, 27–12–2019, 31–01–2020, 16–01–2020
India	17–10–2019, 02–10–2019, 01–11–2019, 11–12–2019, 16–11–2019, 10–01–2020, 26–12–2019, 30–01–2020

TABLE A1.5
Temporal datasets used from Sentinel-1.

Study Site (IW_GRDH)	Sentinel 1A/1B (IW_GRDH)
India	12–12–2019, 11–01–2020, 29–01–2020

TABLE A1.6
Selection of minimum dates using separability analysis on CBSI-NDVI temporal-indices data to identify vegetation type at the Sri Lankan site.

Number of images		Optimum selected dates
Separability value	2	2,3
	3	2,3,6
	4	2,3,4,6
	5	1,2,3,4,6
	6	1,2,3,4,5,6

Note: Dates 1= 06/01/2019, 2=31/01/2019, 3=27/03/2019, 4=27/11/2019, 5=16/01/2020, and 6=31/01/2020.

Table A1.6 shows that maximum-separability values were constant from three images onwards. Hence, dates number 2, 3, and 6 (31/01/2019, 27/03/2019, and 16/01/2020) were selected as the best dates for the first test site (forest). These dates were used to map Rubber and Pinus and secondary forest in the Indikada Mukalana forest. However, Coconut and Para patches were separated using all six dates (06/01/2019, 31/01/2019, 27/03/2019, 27/11/2019, 16/01/2020, and 31/01/2020) because these areas were small and classified images gave better results in place of using just three dates' temporal data. Table A1.7 shows the minimum number of temporal date data for crop-type mapping from the Indian site, using separability analysis on CBSI-NDVI indices temporal images.

From Table A1.7, it can be seen that separability value was constant from five number of images onwards. So, dates 1, 3, 4, 7, and 8 (11/12/2019, 10/01/2020, 30/01/2020, 01/11/2019, and 16/11/2019) for the second test (crops) site were selected as best dates. Minimum dates identified were used to map mustard, wheat, and grass from the Vanasthali area, India. Figure A1.11 shows forest vegetation mapped from first test site. Figures A1.12 and A1.13 show crop fields mapped for the second site with and without SAR images.

The use of CBSI-NDVI images for separability analysis allows for the optimisation of temporal images use to identify distinct tree and shrub vegetation-types in and near forests, as well as for different seasonal crop kinds. It was observed that selected dates (31/01/2019, 27/03/2019, and 16/01/2020) were the best dates to map rubber, pinus, and secondary forest at the Indikada Mukalana forest site in Sri Lanka. But all six dates (06/01/2019, 31/01/2019, 27/03/2019, 27/11/2019, 16/01/2020, and 31/01/2020) had to be used to map small patches of coconut and para, suggesting that

TABLE A1.7
Selection of minimum dates using separability analysis on CBSI-NDVI temporal indices data to identify specific crops type identification at the Indian site.

Number of images		Optimum selected dates
Separability value	1	7
	2	1,7
	3	1,4,7
	4	1,3,7,8
	5	1,3,4,7,8
	6	1,3,4,6,7,8
	7	1,2,3,4,6,7,8
	8	1,2,3,4,5,6,7,8

Note: Number of images 1, 2, 3, 4, 5, 6, 7, and 8 correspond to 11/12/2019, 26/12/2019, 10/01/2020, 30/01/2020, 02/10/2019, 17/10/2019,01/11/2019, and 16/11/2019.

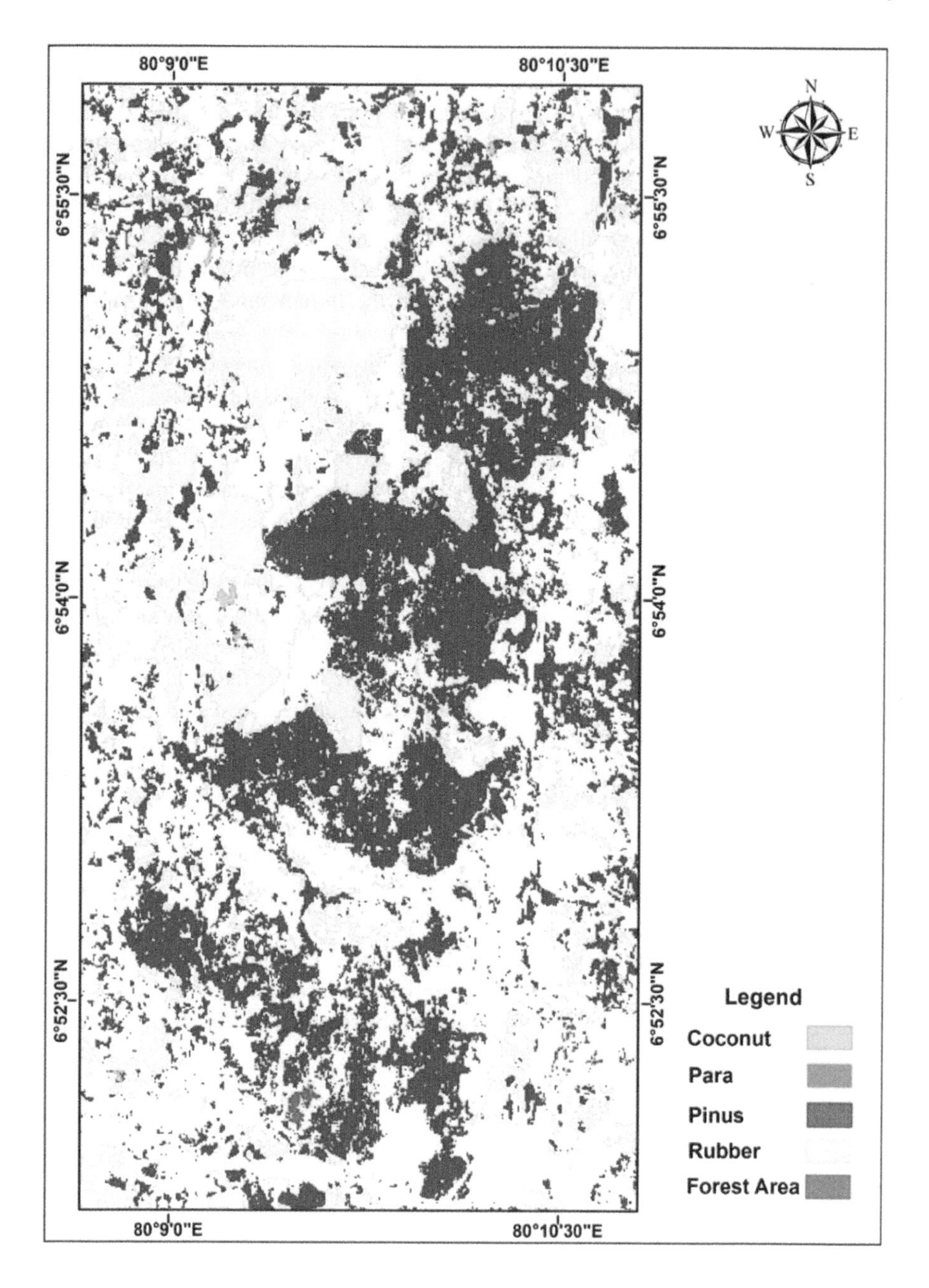

FIGURE A1.11 All forest vegetation-types in Idikada Mukalana area map.

more temporal images gave better results when mapping small patches of a specific vegetation. For the second test site, five temporal dates (11/12/2019, 10/01/2020, 30/01/2020, 01/11/2019, and 16/11/2019) were observed as the best dates to map mustard, wheat, and grass in the Vanasthali region of India.

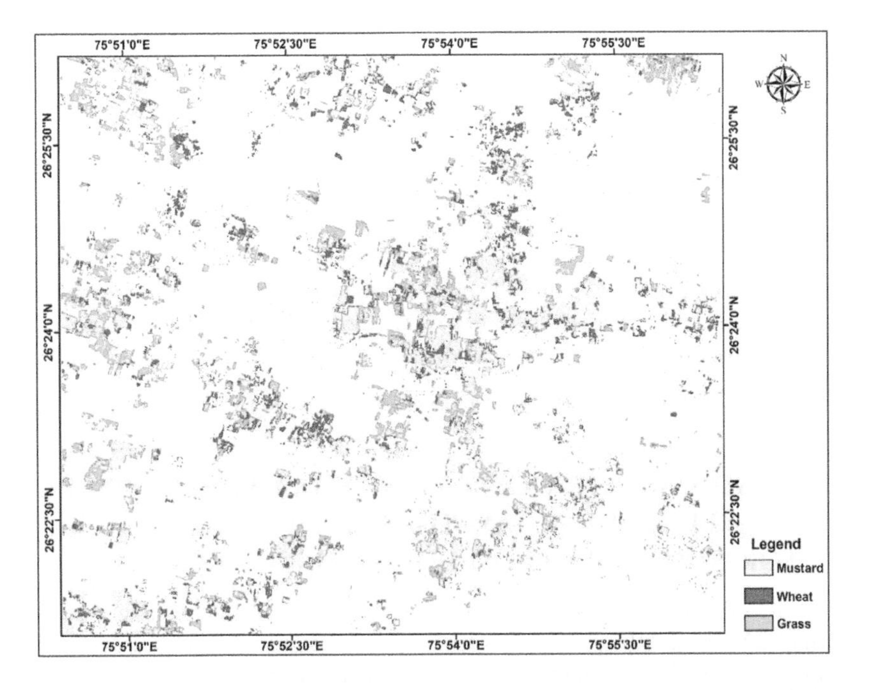

FIGURE A1.12 Mustard, wheat, and grass classified-output map around the Vanasthali area in India.

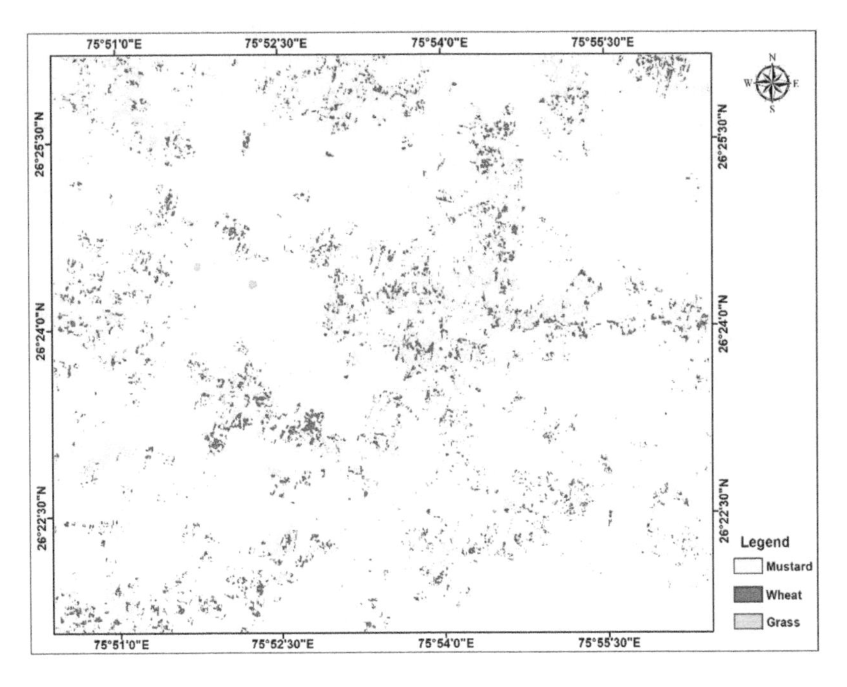

FIGURE A1.13 Mustard, wheat, and grass classified-output map using Sentinel-2 MSI with all Sentinel-1 SAR images.

Specific spectral bands were effective for temporal-indices database-generation as observed in both test site cases. For the Sri Lankan study site, spectral bands red and red-edge-3 bands gave the best CBSI-NDVI values to classify rubber and pinus, and red, blue, and red-edge-3 gave the best CBSI-NDVI values for para and coconut. For the Indian study site, blue, red-edge-2, red-edge-3, and red-edge-4 gave the best CBSI-NDVI values for classifying mustard crops, while red, blue, red-edge-2, and red-edge-3 bands gave the best CBSI-NDVI values for classifying grass.

Second, when optical-temporal remote sensing images were integrated with microwave-temporal images, classification results from combined Sentinel-1 microwave-temporal images with Sentinel-2 optical images were less accurate than Sentinel-2 temporal images.

A3. HANDLING HETEROGENEITY WITH TRAINING SAMPLES USING INDIVIDUAL-SAMPLE-AS-MEAN APPROACH FOR ISABGOL (PSYLLIUM HUSK) MEDICINAL CROP

The temporal images used in this study have bands at different pixel-size from Sentinel-2A/2B satellites. As a part of pre-processing, all the bands were re-sampled to 10m. Vegetation indices approach, using NDVI, three red-edge indices (NDVIRE) and CBSI-NDVI was used to reduce the spectral dimensionality of the temporal images. The mathematical details of these indices is mentioned in Eq. (7.2), Eq. (7.3) and Eq. (7.4) respectively. While considering temporal images representing the complete cycle of the Isabgol crop, separability analysis was done to find out optimum temporal images from temporal-indices database. The optimised images were then considered for Isabgol crop mapping. Training samples were created from training sites and the mean, Variance-co-variance and 'Individual sample as mean' training parameters applied for the mapping of Isabgol crop fields. Mean membership difference (MMD) as a comparison between testing and training fields was done as an accuracy assessment of outputs. MMD should be as low as possible between identical class, whereas it should be high between the class of interest and other classes. The optimised, temporal data-set were used with the 'Individual sample as mean' training approach within the MPCM classifier. From this step, the best vegetation-index approach was identified for mapping the Isabgol crop. Steps in the methodology adopted for this work are mentioned in Figure A1.14.

$$NDVI = \frac{NIR - R}{NIR + R} \tag{7.2}$$

$$NDVI_{RE} = \frac{NIR - RE}{NIR - RE} \tag{7.3}$$

$$CBSI - NDVI = \frac{MAX - MIN}{MAX - MIN} \tag{7.4}$$

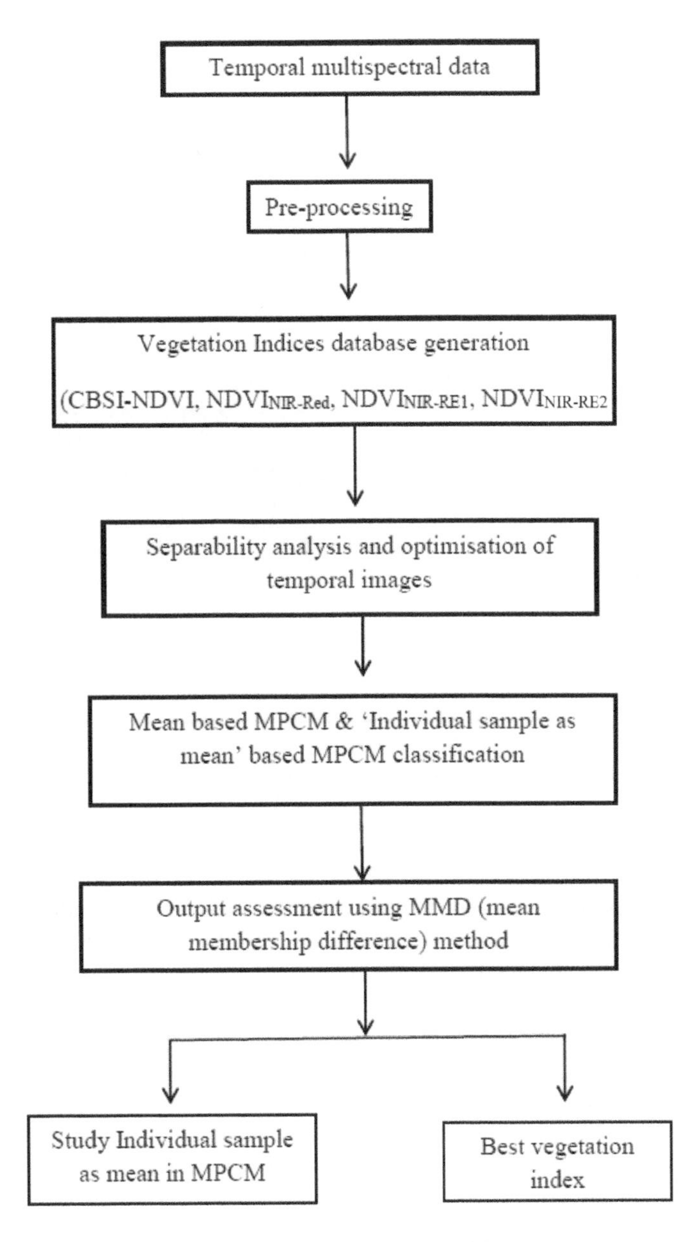

FIGURE A1.14 Methodology adopted for the case study A3.

The study area on which the proposed methodology was tested is Parawa village having 24.8° N latitude and 71.7° E longitude coordinates, situated in Jalore district in the state of Rajasthan, India (Figure A 1.15). Parawa village is extremely far away from its district headquarters, Jalore, with the nearest Sanchore statutory town approximately19kms away. Aspect-wise, Jalore district is towards the south-western

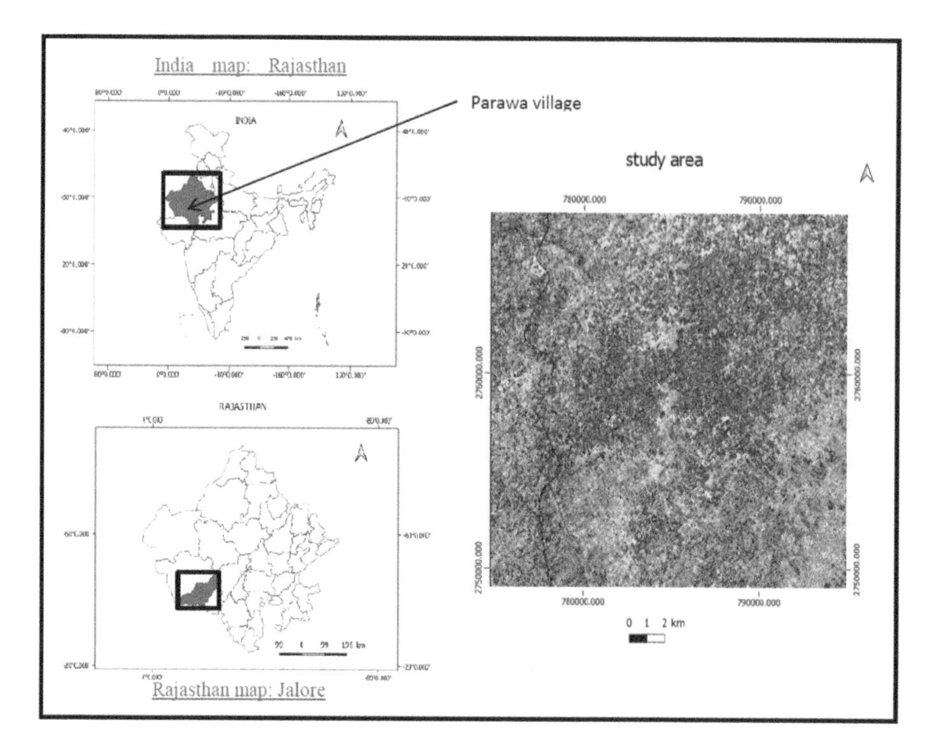

FIGURE A1.15 Location of study area for case study A3.

side of Rajasthan. Barmer is situated towards the north-western border, and Plai is towards the north-eastern side of Jelore, respectively. Sirohiand is to the south-east; and the Gujarat state boundary is to the south.

On 8th January 2021, a ground-truth data-collection field-visit was conducted to collect geo-tagged Isabgol crop fields'. The major agricultural varieties in Parawa village were Isabgol, cumin, mustard, wheat, castor plant, and fenugreek. Different crop fields were identified through geo-tagged GNSS observation coordinates and captured by field photographs during the visit. Training fields for each of the crops are marked on the Google Earth image as shown in Figure A1.16. The field photographs are shown in Figures A1.17 to A1.20.

The goal of employing temporal Sentinel-2A/2B data for this research study was to have a five-day temporal resolution, which assisted in catching unique stages of Isabgol (Delegido et al., 2011).

Table A1.8 shows satellite data used in the research work whereas table A1.9 depicts different phenological growth stages of ishabgol crop. As per the Isabgol crop cycle, Sentinel-2 satellite images from December 2020 to March 2021, covering the whole crop cycle and having temporal gaps of approximately 10 days were used in this research work. Temporal dates from Sentinel-2 satellite covered in this research work were 31st December 2020, 10th January 2021, 04th February

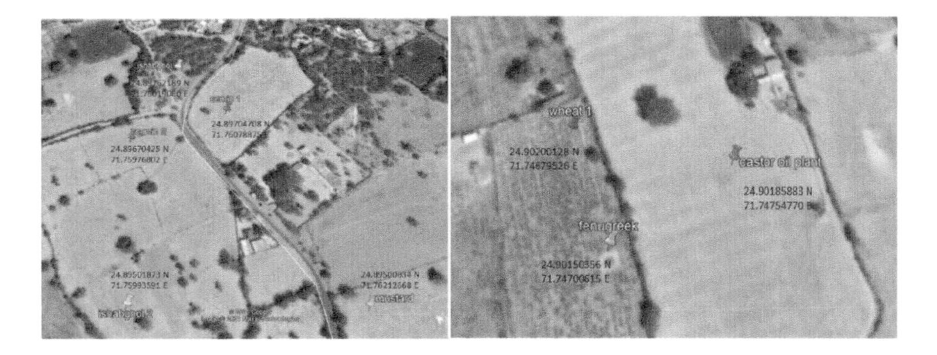

FIGURE A1.16 Training fields marked on Google Earth images of various crops (Parawa village, Rajasthan).

FIGURE A1.17 Isabgol field photographs (Parawa village, Rajasthan).

FIGURE A1.18 Mustard field photographs and cumin field photographs (Parawa village, Rajasthan).

FIGURE A1.19 Wheat field photographs and fenugreek field photographs (Parawa village, Rajasthan).

FIGURE A1.20 Castor field photographs (Parawa village, Rajasthan).

TABLE A1.8
Data details used for this study area of Sentinel 2A/2B.

Study site	Sentinel 2A data dates	Sentinel 2B data dates
Parawa village, Rajasthan	04 February 2021	31 December 2020
	14 February 2021	10 January 2021
	16 March 2021	01 March 2021
	26 March 2021	21 March 2021

2021, 14th February 2021, 1st March 2021, 16th March 2021, 21st March 2021, and 26th March 2021.

From Table A1.9, based on temporal-optimised dates of CBSI-NDVI indices—e.g. 10th January 2021, 14th February 2021, 1st March 2021, 16th March 2021, and 26th March 2021—growth stages of Isabgol considered were budding, pollination, ripening, seed maturing, and sun drying. In the same way, based on temporal-optimised dates of NDVI indices—e.g. 10th January 2021, 1st March 2021, 16th March 2021, 21st March 2021, and 26th March 2021—growth stages of Isabgol considered were flowering, seed maturing, harvesting, and psyllium-crop sun drying. From temporal-optimised dates of NDVINRE1 (NIR red-edge-1) indices, 10th January 2021 represents flowering; 1st March 2021 represents seed maturing; 16th March 2021 represents harvesting; and 1st March 2021 and 26th March 2021 show the sun-drying stage. Based on temporal-optimised dates of NDVINRE2 indices studied, such as 31st December 2020, 10th January 2021, 4th February 2021, and 21st March 2021, Isabgol crop stages found were budding, flowering, pollination, and plant sun-drying. Furthermore, based on temporal-optimised dates of NDVINRE3—e.g. 10th January 2021, 4th February 2021, 14th February 2021, 1st March 2021, and 26th March 2021, corresponding growth stages found were flowering, pollination, ripening, seed maturing, and plant sun-drying.

From pre-processing steps of separability using temporal-indices input data, temporal image dates can be optimised. A temporal-indices database can be generated from conventional mathematical formulas or from the CBSI-NDVI approach. For Isabgol crop-mapping, following optimised-temporal-image data gave best classification results, which were as follows: 10th January 2021, 14th February 2021, 1st March 2021, 16th March 2021, and 26th March 2021. Based on the optimised-temporal-image indices database, growth stages of Isabgol

TABLE A1.9
Isabgol crop-growth stages as per satellite images used.

Approximate	Growth stage
31/12/2020	Seedling
10/01/2021	Budding
04/02/2021	Flowering
14/02/2021	Pollination
01/03/2021	Ripening
16/03/2021	Seed maturing (yellowing of leaves)
21/03/2021	Harvesting
26/03/2021	Sun drying of plant

crop represented through temporal images can be known. Main growth stages of the Isabgol crop represented through optimised-temporal images were flowering, ripening, seed maturing, harvesting, and plant sun-drying. Various forms of generated indices were CBSI-NDVI, NDVI (Red-NIR), NDVINRE1 as NIR red-edge-1, NDVINRE2 as NIR red-edge-2, and NDVINRE3 as NIR red-edge-3. Temporal output from these indices methods were used as input in the fuzzy MPCM classifier. The fuzzy MPCM classifier was being used as a supervised-classifier approach. So, three different types of training-parameter approaches were tested in this research work for mapping the medicinal crop Isabgol. These three training-parameter approaches were mean and variance-covariance-based MPCM, conventional mean-based MPCM, and 'individual sample as mean' based MPCM (Figure A1.21).

It has been observed that mean and variance co-variance-based training approaches overestimated the Isabgol fields, and secondly, heterogeneity within fields were not handled. So, mean and variance co-variance-based training approaches were not considered for further analysis. 'Individual sample as mean' innovative-training-parameters approach in MPCM classifiers handled heterogeneity within classes while mapping Isabgol crop fields with high accuracy. Accuracy assessment of mapped Isabgol fields was conducted using the MMD method. It was observed with respect to different indices that different crops gave the maximum MMD value with Isabgol crops. These crops verses indices combinations are, for example: NDVI (Red-NIR) for wheat; NDVINRE1 (NIR red-edge-1) for mustard and castor; NDVINRE3 (NIR red-edge-3) for fenugreek;

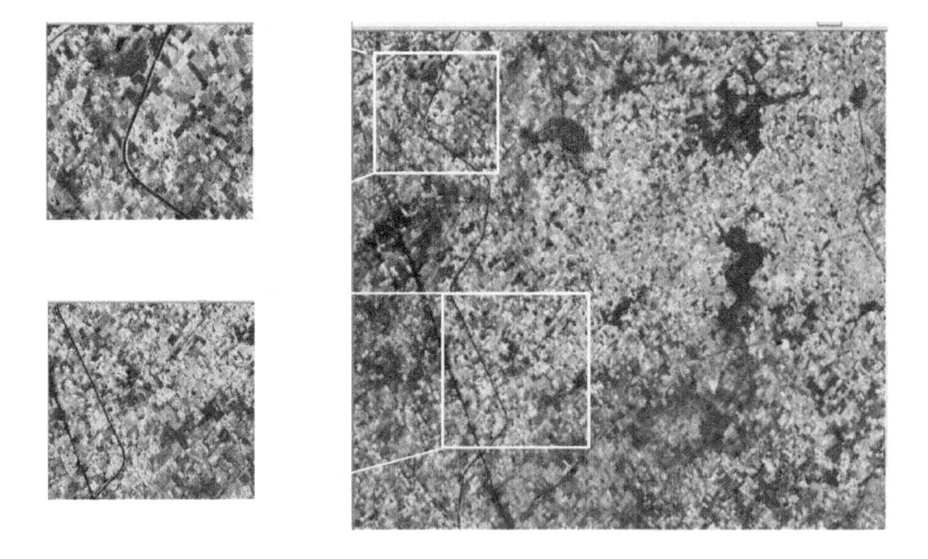

FIGURE A1.21 Isabgol crop fields identified from temporal CBSI-NDVI indices database while applying an 'Individual sample as mean' based MPCM classifier.

and CBSI-NDVI for cumin. It was observed that the least MMD was observed with fenugreek. That means the fenugreek crop was very spectrally close with the Isabgol crop. Band combinations used through the CBSI-NDVI approach for different temporal images were: minimum as the red band was blue (490 nm) band; maximum as the NIR band in different temporal images were red-edge (865nm), NIR (842nm), and SWIR (1375 nm). Overall, the CBSI-NDVI temporal database gave the best result with minimum outliers. This proposed methodology has wider applications in specific-crop mapping and studying various plant phenologies.

A4. SUNFLOWER CROP MAPPING USING FUZZY CLASSIFICATION WHILE STUDYING EFFECT OF RED-EDGE BANDS

In this research work, sunflower crop-fields have been mapped from the Shahabad area of Kurukshetra district, Haryana, India using temporal Sentinel-2 datasets. So, in all, 20m bands were re-sampled to 10m as a pre-processing step. This resampled temporal data of 10m was used as input for the generation of a temporal-indices database. Various vegetation indices considered were, for example, SR, NDVI, CI, and SAVI while considering various proportions of red edge and red bands. The temporal-indices database was fed as the input for the classifier. MPCM classification was applied for a particular range of membership-weighting components or fuzzy weight-constants (m). The output assessment was conducted through MMD method to identify the best vegetation index, suitable red-edge bands, and corresponding weight-estimation with optimised temporal images for different sowing periods. The methodology adopted for this research work is shown in Figure A1.22.

STUDY AREA

Shahabad tehsil of Kurukshetra district of Haryana state, India has been selected as the study area (Figure A1.23). Wheat as well as rice production is major in the north-central part of India, especially in the state of Haryana. Shahabad is located with central coordinate 30° 10' 6" North, 76° 52' 12" East and is between Ambala and Kurukshetra, two major cities having fertile agricultural land. The major crops cultivated are sunflower, wheat, and vegetables. This case study was carried out during the Rabi season of Indian cropping system.

For the identification of sunflower-crop training and testing fields on 10th April 2019, a field visit was conducted. The major agriculture varieties during Rabi season in this study site are sunflower and wheat. In the three villages of Shahabad tehsil, named Nalvi, Kalsana, and Jharauli, sunflower-crop production was found to be abundant. During a field visit, three different stages of sowing of sunflower fields viz., early-sown, middle-sown, and late-sown. During the field visit, it was observed that the wheat crop was at the harvesting period with a few exception. Images of the sunflower fields taken during the field visit are shown in Figures A1.24, A1.25, and A1.26.

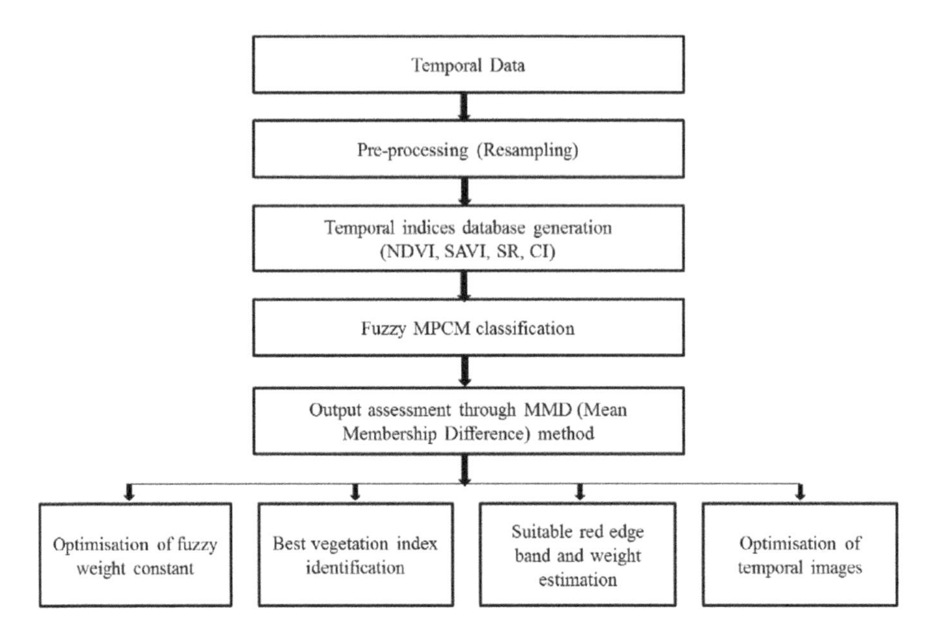

FIGURE A1.22 The methodology adopted for sunflower-crop mapping in case study A4.

Based on field visit conducted, it was found that temporal images dated from 21st March 2019 onwards were suitable for mapping sowing-period information of early-sown sunflower fields. Temporal data mentioned in Table A1.10 were used in the study.

The major purpose of this study was to explore the impact of red-edge bands on sunflower-crop mapping using a fuzzy-classification approach. The objectives defined for the study are listed as follows:

a) Identification of a suitable vegetation-index method to generate a temporal-indices database.
b) Red-edge band suitability with weightage assigned to the red-edge bands to be estimated.
c) Identification of temporal remote-sensing images for early-sown, middle-sown and late-sown sunflower fields.

To achieve the aforementioned goals, temporal datasets from the Sentinel-2 MSI sensor in three red-edge bands from 21st March 2019 were examined for sunflower-crop field mapping in the Shahabad area of Haryana, India. During the field investigation, it is recorded that the sunflower crop is cultivated in three different sowing periods (Figures A1.24 to A1.26). NDVI, SAVI, SR and CI modified indices were used to generate a temporal-indices database. To handle noise, coincident-cluster problems, outliers, and single-class extraction effectively, MPCM classification technique was applied. Mean Membership Difference (MMD) between fields of the same sunflower-sown fields or between sunflower fields and wheat fields were considered for an indirect accuracy-assessment. MMD values within sunflower fields should tend to be zero, whereas values between wheat and sunflower fields should tend to be one.

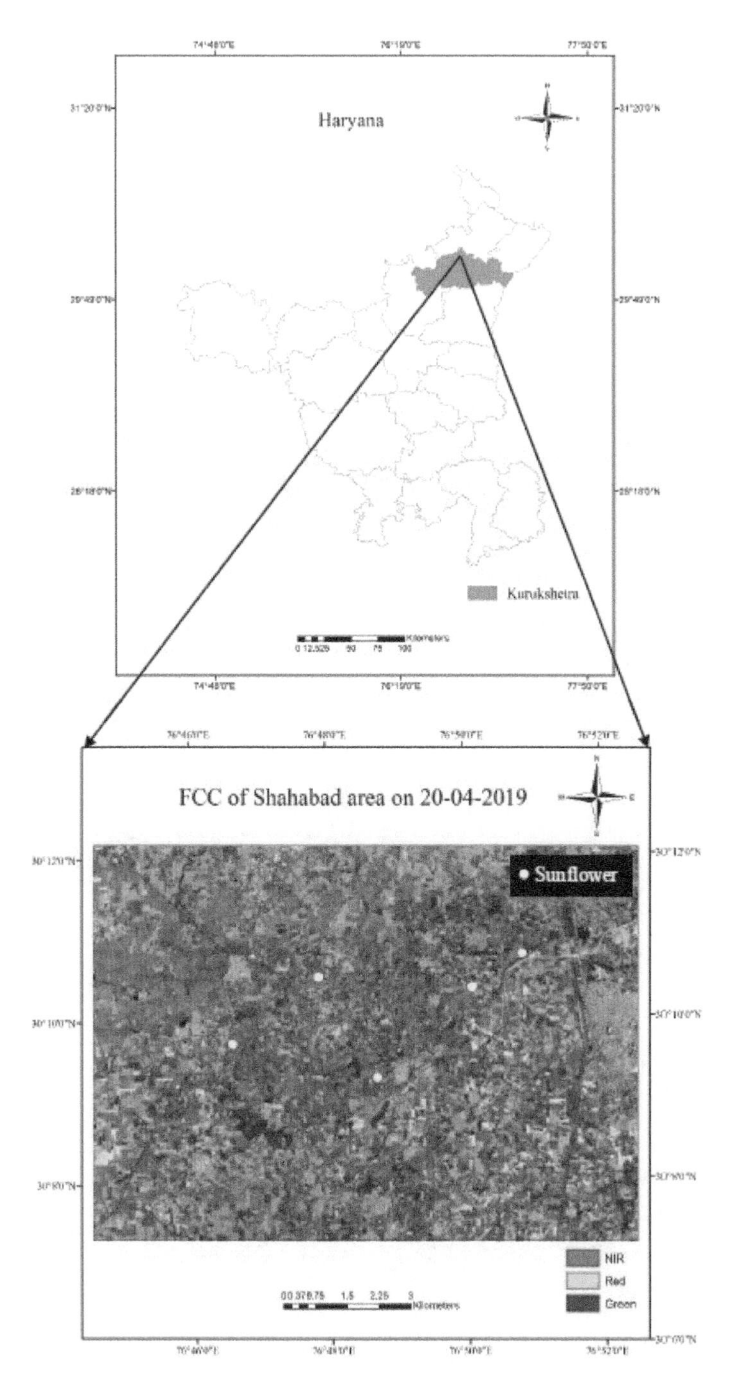

FIGURE A1.23 Study area: Shahabad, Haryana, India.

FIGURE A1.24 Early-sown sunflower field in Nalvi village.

FIGURE A1.25 Middle-sown sunflower field near Shahabad Markanda.

FIGURE A1.26 Late-sown and early-sown sunflower fields near Jharouli village.

TABLE A1.10
Temporal datasets used in this research work.

Sentinel 2A (L2A product)	Sentinel 2B (L2A product)
21/03/2019	26/03/2019
31/03/2019	05/04/2019
20/04/2019	15/04/2019
30/04/2019	05/05/2019

During the classification of sunflower crop in the selected area, it has been found that the degree of fuzziness was independent in the range from 1.2 to 3, and hence, this established the homogeneity of the crop fields considered. It has also been found that at 'a' = 0.7 the optimised weight assigned to red-edge band for the calculation of modified vegetation-indices.

It has also been observed that temporal-indices databases generated through modified CI gave the best results in separating sunflower crops from wheat fields, while temporal databases generated through the SAVI method gave the best results in mapping sunflower-crop sowing stages at different times.

It has been observed that different red-edge bands were suitable with different indices while discriminating and mapping sunflower-crop stages with wheat crops. Red-edge band-1 has been found suitable with NDVI; red-edge band-2 and 3 have been found suitable with SAVI for sunflower-crop stage mapping with wheat crop. Red-edge band-2 has been found suitable with SR as well as CI for accurately mapping sunflower-crop fields. Red-edge band-1 in a modified, SAVI-based temporal-indices database have given very good result for mapping early-sown, middle-sown, and late-sown sunflower-crop fields while other red-edge bands used with conventional indices were less effective in mapping various stages of sunflower-crop fields.

The graph between temporal-indices values of these crop-stage fields and temporal dates was used to identify optimal temporal pictures for early sown, middle sown, and late sown. Temporal-indices values of sunflower-crop fields were considered from fields identified during a field visit. While using CI indices, identification of early-sown sunflower fields were using red-edge band-2 with weight value ('a' = 0.7) for temporal dates of 21st March 2019, 26th March 2019, 31st March 2019, and 15th April 2019. Middle-sown sunflower fields were identified nicely using red-edge band-2 with weight value of ('a' = 0.7) while using images of temporal dates like 26th March 2019, 31st March 2019, 5th April 2019, and 15th April 2019. Late-sown sunflower fields were identified better while using red-edge band-1 with weight value ('a' = 0.7) for 31st March 2019, 5th April 2019, and 15th April 2019. SAVI indices with red-edge 1 yielded the same temporal dates as CI indices for early-sown, middle-sown, and late-sown sunflower-crop fields. Figure A1.27 depicts the best results obtained while applying modified CI indices.

From the case study, it is quite clear that using red-edge bands in indices gave much better results for mapping sunflower crop through MPCM classification. But mapping different sowing stages of sunflower crops with red-edge bands was less effective.

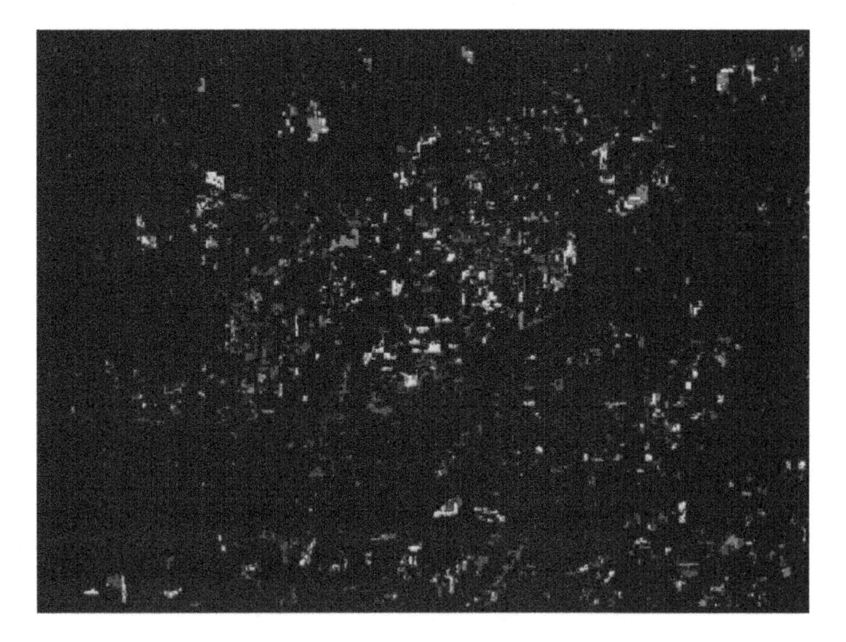

FIGURE A1.27 Different sowing stages of a sunflower crop while using modified CI indices temporal database.

A5. MAPPING BURNT PADDY FIELDS USING TWO DATES' TEMPORAL SENTINEL-2 DATA

Mapping of stubble burnt paddy fields in surrounding areas of Patiala City, India was conducted using an optimised MPCM algorithm. The burning of paddy stubble in the northern part of India is one of the most common practices. Sentinel-2A and Sentinel-2B temporal dataset was considered for the mapping of stubble burnt paddy fields from 10th Oct 2019 to 19th Nov 2019 within intervals of two to five days. The extraction of burning patches was done for both KMPCM and KMPCM-S. The CBSI approach was used to reduce the dimensionality and classify the burnt patches. NBR was used for cross-validation and it was observed that paddy stubble burning increases rapidly after 20th Oct 2019 and was at its peak by 4th Nov 2019. The field visit was conducted for collecting geo-tagged burnt paddy fields sample on 24th Oct 2019 and 25th Oct 2019. Collected geo-tagged training samples in the course of the field visit is shown in Figure A1.28 (Chhapariya et al., 2021a, 2021b, 2020).

The study region's geographic coordinate border was 30°26'23.76" N, 76°5'30.44" E to 30°10'0.42" N, 76°23'58.11" E with an area of 30km*30km and is located in the district of Patiala, Punjab state, India (Figure A1.29). The identified area mainly has a paddy crop in Kharif session, as well as a eucalyptus plantation and a potato crop by the end of the session of Kharif crops. The following was the primary reason for selecting this research topic: to put the optimised kernel-based MPCM method for testing. During October to November months there is large-scale burning of

FIGURE A1.28 During 24 Oct 2019 and 25 Oct 2019, collected GNSS-linked stubble burnt paddy field in Patiala district.

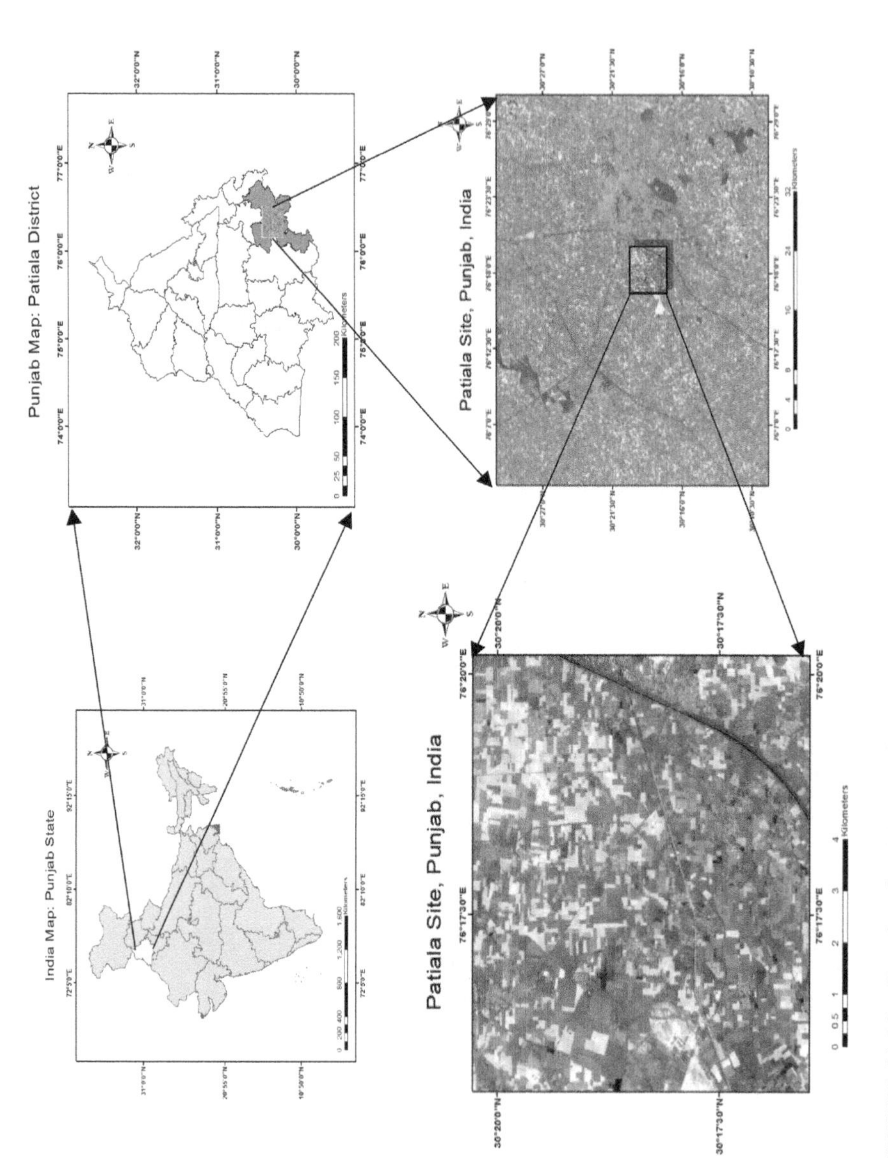

FIGURE A1.29 Test site of paddy burnt fields.

paddy stubble in Haryana and Punjab state, which contributes heavy pollution in NCR (National Capital Region). Secondly, throughout the months of October and November of 2019, there was cloud-free optical Sentinel-2 data available.

As part of the process for this case study, all required bands with a spatial resolution of 20m were re-sampled to 10m, and then Band-2, Band-3, Band-4, Band-8, Band-11, and Band-12 were layer stacked. Using information from ground-truth samples, training-sample data for burnt paddy fields were developed. Further cognitive science concepts include historical knowledge used to identify training samples needed for future temporal images, as well as temporal images utilised in the previous few days to map stubble burnt rice fields. The MPCM algorithm objective function for distance, as well as kernel (Gaussian kernel)-based with optimised parameters was used. Mapping of the stubble burnt paddy field was done every two to five days, depending on the availability of temporal images. Delta Normalised Burnt Ratio (dNBR) and Normalised Burnt Ratio (NBR) were used for validating the results (Figure A1.30).

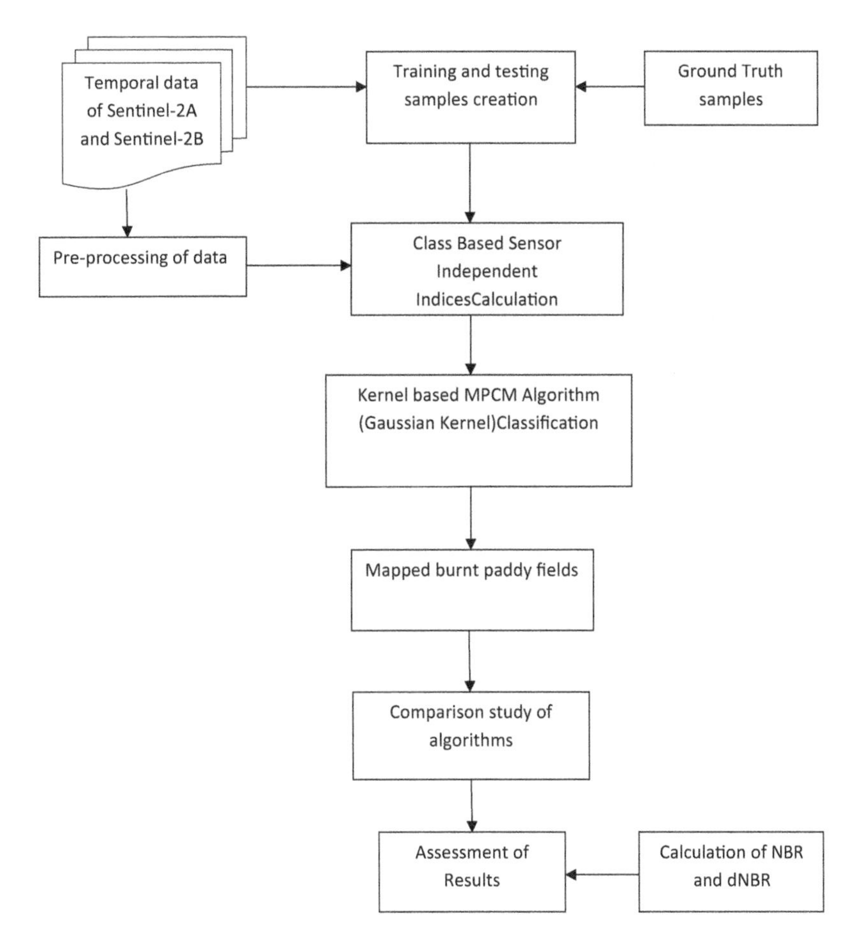

FIGURE A1.30 Research methodology (mapping of burnt paddy field).

An optimised KMPCM algorithm was used for mapping burnt paddy fields as a single class. A CBSI approach was used to reduce the dimensionality of multispectral two-date temporal-images. The burnt patches were mapped from 10th Oct 2019 to 19th Nov 2019 while using kernels-based MPCM with and without spatial information. For KMPCM (Table A1.11) and KMPCM-S (Table A1.12), burnt area and number of burnt pixels were computed for each date. Normalised Burnt Ratio (NBR) was used for the cross-validation of results. NBR has been compared with classified as well as CBSI output for both KMPCM (Figure A1.31) and KMPCM-S (Figure A1.32). The

TABLE A1.11
Burnt area and burnt pixel for different dates using KMPCM.

Acquisition date	Burnt pixels	Burnt area in sq. km
10th Oct 2019	11790	1.1
15th Oct 2019	19316	1.9
17th Oct 2019	24106	2.4
20th Oct 2019	43553	4.3
22nd Oct 2019	183269	18.3
25th Oct 2019	298029	29.8
27th Oct 2019	302489	30.2
4th Nov 2019	465969	46.5
6th Nov 2019	439153	43.9
9th Nov 2019	895749	89.5
19th Nov 2019	693368	69.3

TABLE A1.12
Burnt pixel and burnt area for different dates of acquisition for KMPCM-S.

Acquisition Date	Burnt pixels	Burnt area in sq. km
10th Oct 2019	10621	1.0
15th Oct 2019	15909	1.5
17th Oct 2019	19178	1.9
20th Oct 2019	32764	3.2
22nd Oct 2019	127453	12.7
25th Oct 2019	249841	24.9
27th Oct 2019	267775	26.7
4th Nov 2019	358423	35.8
6th Nov 2019	335275	33.2
9th Nov 2019	612347	61.2

FIGURE A1.31 MPCM classified output and NBR classified output.

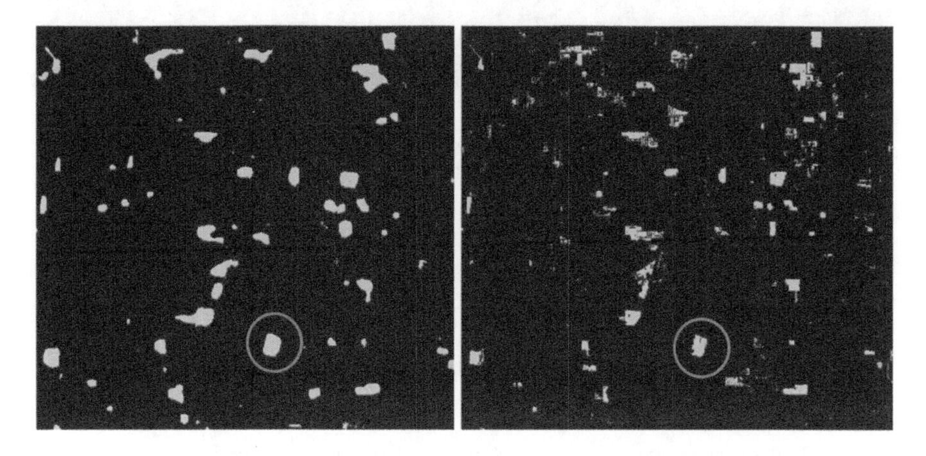

FIGURE A1.32 Cross-validation of results with and without spatial constraint of classified output on 25th October.

NBR output for other advanced base classifiers with spatial constraint is shown in Figure A1.33.

Mean Membership Difference (MMD) method has been used as a quantitative assessment of mapped paddy stubble burnt fields from soft-classified outputs. The accuracy assessment had been done for kernel-based MPCM algorithm outputs with and without spatial information. From the results, it is clear that the KMPCM algorithm with spatial constraints has shown better performance while using the MMD accuracy-assessment method (Table A1.13 and Table A1.14).

FIGURE A1.33 Cross-validation of results with different spatial constraints and advance-base classifier of classified output on 25th October 2019.

TABLE A1.13
Accuracy assessments for KMPCM algorithm without spatial constraint using MMD.

Date of acquisition	KMPCM without spatial constraint	
	MMD (Burnt patches)	MMD (Unburnt patches)
10th Oct 2019	0.0807	0.4509
15th Oct 2019	0.0721	0.5345
17th Oct 2019	0.0812	0.5012
20th Oct 2019	0.0457	0.6481
22nd Oct 2019	0.0398	0.762
25th Oct 2019	0.012	0.8912
27th Oct 2019	0.019	0.9438
4th Nov 2019	0.0342	0.7291
6th Nov 2019	0.0253	0.7981
9th Nov 2019	0.0637	0.5902
19th Nov 2019	0.0445	0.6834

TABLE A1.14
Accuracy assessments for KMPCM algorithm with spatial constraint using MMD.

Date of acquisition	KMPCM with spatial constraint	
	MMD (Burnt patches)	MMD (Unburnt patches)
Oct 10 2019	0.0307	0.7509
Oct 15 2019	0.0321	0.8345
Oct 17 2019	0.0212	0.7012
Oct 20 2019	0.0157	0.7481
Oct 22 2019	0.0198	0.8623
Oct 25 2019	0.009	0.9912
Oct 27 2019	0.012	0.9438
Nov 4 2019	0.0242	0.8241
Nov 6 2019	0.0153	0.7967
Nov 9 2019	0.0237	0.8802
Nov 19 2019	0.0345	0.7534

A6. MAPPING TEN-YEAR-OLD DALBERGIA SISSOO FOREST SPECIES

Information on species mapping is a critical approach for forest management and long-term conservation practises. Remote-sensing data has proven to be an asset for the assessment of the spatial distribution of species over time. This study mapped Dalbergia sissoo, a timber species found on both sides of the Jakhan River in the Dehradun district of the lesser Himalayas, using time-series data from a single sensor (PlanetScope) and a dual-sensor (PlanetScope with Sentinel 2). A part of the Barkot Forest Range and its surroundings were selected for the study. The study area is a Sal-Mixed, Moist, Deciduous Forest and is bounded by the Jakhan River in the west, the Ganga River in the east, the Thano Forest Range in the north, and the Motichur Forest Range in the south. Its terrain ranges from flat to undulating. Sal (*Shorearobusta*) is the dominant tree species in the area. The co-dominant species are Dhaura (*Lagerstroemia parviflora*), Sain (*Terminalia tomentosa*), and Jamun (*Syzygiumcumini*), among others. Teak (*Tectona grandis*), Shisham (*Dalbergia sissoo*), Khair (*Acacia catechu*), and Domsal (*Miliusavelutina)* are some of the other species found in the area. The study area has patches of *Dalbergia sissoo* in association with *Acacia catechu* growing along the Jakhan River. This study focused on the extraction of Shisham growing in the Jakhan Riverbed (Figure A1.34).

The different phenological stages of Shisham based on literature review and field survey are given in Table A1.15.

Segregation of trees based on age can be advantageous for classification, as it can affect the spectral response and the interclass variability of the trees. Three distinct classes of the target species were observed in the field based on their height and growth stage (Table A1.16). These classes were "young Shisham 1" (YS1), "young

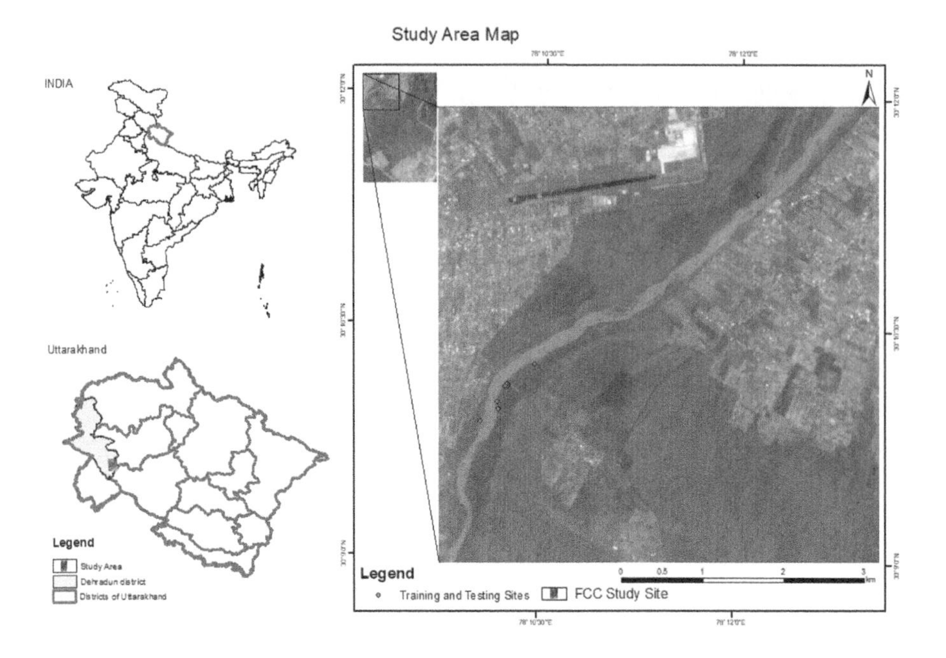

FIGURE A1.34 Location of the study area for case study A6.

TABLE A1.15
Phenological stages of *Dalbergia sissoo*.

Phenological stage	Months
Leaf fall	October-February
Leaf flush	February-April
Flowering	March-April
Fruiting	April-May (almost for the whole year)
Seed formation	July-December

TABLE A1.16
Description of distinct classes of Shisham observed in January 2022 in the study area.

S. No.	Class	Age Group	Height	Growth Stage
1	YS1	<5 years	1–5m	Completely leafless
2	YS2	6–10 years	5–10 m	Few leaves
3	MS	>10 years	>10 m	More leaves than YS2

FIGURE A1.35 Three categories of Shisham found in the study area in January 2022 (a-c). (a) Young Shisham 1 (YS1); (b) Young Shisham 2 (YS2); (c) Mature Shisham (MS); (d) shows a patch of YS2 in May 2022 satellite data.

Shisham 2" (YS2), and "matured Shisham" (MS) (Figure A1.35). Young Shisham 1 had completely dry undergrowth, whereas the other two classes had relatively less dry and shrubby undergrowth. This study focused on extracting the second class—i.e. "young Shisham 2" (YS2).

The *PlanetScope* offers data at a spatial resolution of 3m on daily basis. It gives the data in blue, green, red, and near-infrared bands. For the current study, the 4-band surface-reflectance products with a cloud cover of less than 10% were selected. Twenty-one images of different months of the year 2021 were selected based on availability and their relevance with the phenological stage information of *Dalbergia sissoo*. The dates selected are given in Table A1.17.

A constellation of Sentinel-2A/2B satellites offers 13-band data at an interval of five days. For the current study, level-2A products with a cloud cover of less than 10% were selected. Three images, one each from March, April, and May, were used to increase the number of dates in the transition phase of leaf fall and leaf flush of *Dalbergia sissoo* and test the compatibility of Sentinel-2 data with *PlanetScope* data. The dates used for the study are given in Table A1.18.

TABLE A1.17
List of dates of *PlanetScope* 4-band images used in the study along with their corresponding phenological stages.

S. No.	Date	Phenological stage
1	09/01/2021	Leaf fall
2	31/01/2021	
3	08/02/2021	Leaf fall-flush
4	17/02/2021	Leaf flush
5	28/02/2021	
6	11/03/2021	Leaf flush-flowering
7	08/04/2021	Leaf flush flowering—fruiting
8	09/05/2021	Fruiting
9	14/05/2021	
10	08/06/2021	
11	01/07/2021	Fruiting—seed formation
12	27/09/2021	
13	09/10/2021	Fruiting—seed formation—leaf fall
14	16/10/2021	
15	30/10/2021	
16	02/11/2021	
17	08/11/2021	
18	17/11/2021	
19	04/12/2021	
20	11/12/2021	
21	31/12/2021	

TABLE A1.18
Dates of *Sentinel-2* images.

S. No.	Date	Phenological Stage
1	25/03/2021	Leaf flush- flowering
2	24/04/2021	Leaf flush—flowering—fruiting
3	24/05/2021	Fruiting

Methodology adopted is shown in Figure A1.36.

The NDVI classified outputs of the single and the combined dataset are shown in Figures A1.37(a) and A1.37(c), respectively. It was again observed that the results with single-sensor data were more specific than that with the combined sensor output. The single-sensor data set was able to detect more pixels of the target class

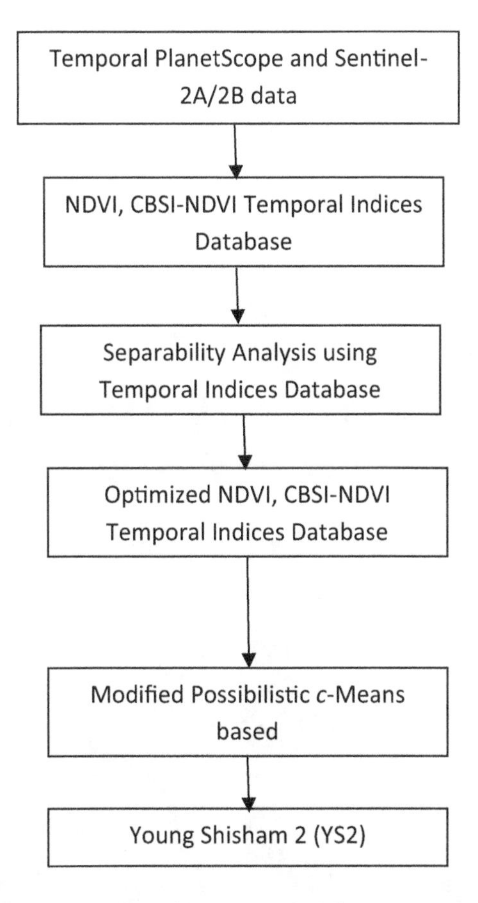

FIGURE A1.36 Methodology adopted in case study A6.

in comparison with the combined sensor-output. This could be due to the higher spatial-resolution *PlanetScope* dataset. Figures A1.37(b) and A1.37(d) depict the CBSI-NDVI output. This output revealed that the target class was less identifiable using both single-sensor and dual-sensor approaches when compared to the NDVI technique.

The study is a preliminary investigation of *PlanetScope* 4-band data and combined data of *PlanetScope* and Sentinel-2 for single-species extraction. This study used single-sensor and dual-sensor approaches to test two indices; namely, NDVI and CBSI-NDVI, with the MPCM algorithm and two training-parameter concepts; namely, mean and novel ISM approach. The study showed that Euclidean distance gave the best measure of separability for both NDVI and CBSI-NDVI approaches. It was found that CBSI-NDVI tested with the ISM approach of MPCM gave slightly better results than NDVI with the same approach when tested using MMD.

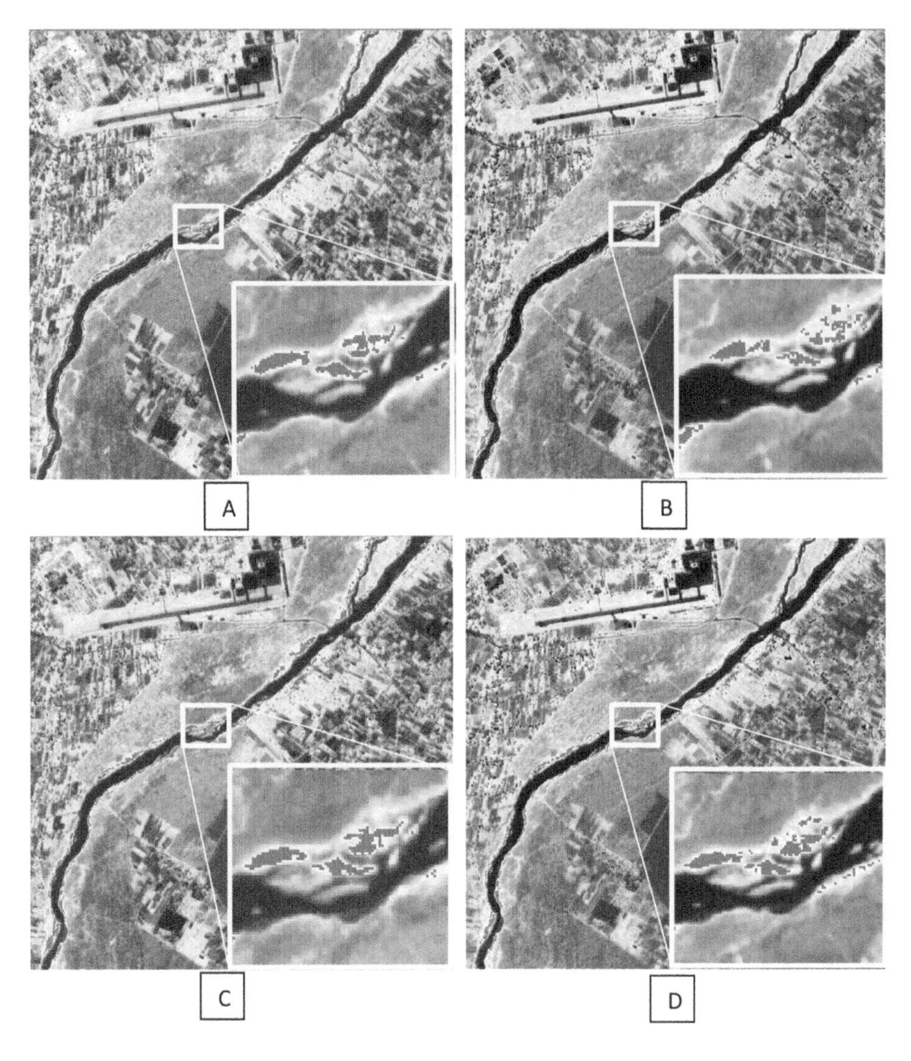

FIGURE A1.37 Results of MPCM with ISM approach. Figures (a) and (b) are single-sensor output with NDVI and CBSI-NDVI, respectively; Figures (c) and (d) are combined-sensor output with NDVI and CBSI-NDVI, respectively. The yellow inset shows an enlarged section from the study site for better visualisation.

A7. TRANSITION BUILDING FOOTPRINTS MAPPING

This study presents a fuzzy approach for the detection of transitioned building-footprints in urban areas using medium-resolution data sets. Multi-temporal remote-sensing data sets from Landsat-8 Operational Land Imager and Sentinel-2A were used for generation of a temporal-indices database. The database was generated using a CBSI-NDVI, with an aim to reduce spectral dimensionality of each image and maintain temporal dimensionality. The temporal-indices database was subsequently used as input in MPCM Classifier for transitioned building-footprints extraction. The identified transitioned-building locations were validated using ground samples, as well as from Google images at four different test sites. For accuracy assessment, F-measure was calculated and its value found to be 0.75 or higher for all training and testing sites. Thus, using a proposed fuzzy approach, transitioned-building footprints were accurately identified compared to traditional techniques.

Dehradun City, located in the Uttarakhand state of India, is the study area for the research work. The city has a geographical extent of 30° 15′ 18.362″ N to 30° 20′ 54.693″ N and 77° 58′ 1.464″ E to 78° 6′ 13.978″ E. Dehradun is the primary city in the region and the neighbouring towns and settlements have a symbiotic relationship with Dehradun City, depending on it for higher-level services. The city has thus emerged as a vital service centre, since the trade and commerce requirements of the region and higher-order facilities of health, education, recreation, and transportation of the surrounding hinterland are met by the city. As a result, during the last few decades, the city has registered an unprecedented sprawl in its area and is expanding rapidly onto the surrounding fertile agricultural lands. These developmental pressures have stressed Dehradun city to the breaking point, besides disturbing various hydrological and ecological cycles and causing the loss of arable land. Figure A1.38 shows the key-map for the study area. For training the algorithm, 50 samples were collected from site 1 (Bharuwala Colony as training site). As per Jensen (Jensen, 1986), the number of training samples per class should be 10n, where 'n' is dimensionality of data. In the present study, since a two-dimensional indices database was taken as input to the MPCM classifier, the value of n = 2. However, we took the training samples to be 50 instead of 20, and to have a better representation of transitioned building footprints. Seventy-five testing samples were collected from four testing sites (Indian Institute of Remote sensing [IIRS] Kalidas Road, Doon University area, Sabji Mandi area, and ISBT as testing sites). As per Congalton (Congalton, 1991), 75–100 testing samples per class are sufficient; that's why 75 testing samples were collected from four different sites.

Images acquired by operational land-imager (OLI) sensor on board Landsat-8 and multi-spectral instrument (MSI) sensor on board Sentinel-2A (Table A1.19) were used in the study. The Landsat-8 sensor data was acquired on 20–10–2013 (T1) and Sentinel-2A on 22–10–2019 (T2). The time difference between the two datasets was taken to be six years to ensure that a sufficient number of transitioned-buildings footprints from open spaces are available. The Landsat-8 multi-spectral data was merged with panchromatic data of Landsat-8 satellite using a wavelet-fusion algorithm (Nünez et al., 1999; Sokolova et al., 2006). The wavelet-based approach generated fused data at 15m, which was subsequently resampled at 10m to ensure compatibility with Sentinel-2A data.

Small transitioned-building footprints could not be extracted as the first image (T1) is of coarser resolution (i.e. 30 m). Secondly, if the images (of T1 and T2) were from the same sensor, it would result in improved image-to-image registration, which would, again, result in better pixel-to-pixel correspondence between the two temporal images and eventually may lead to more accurate

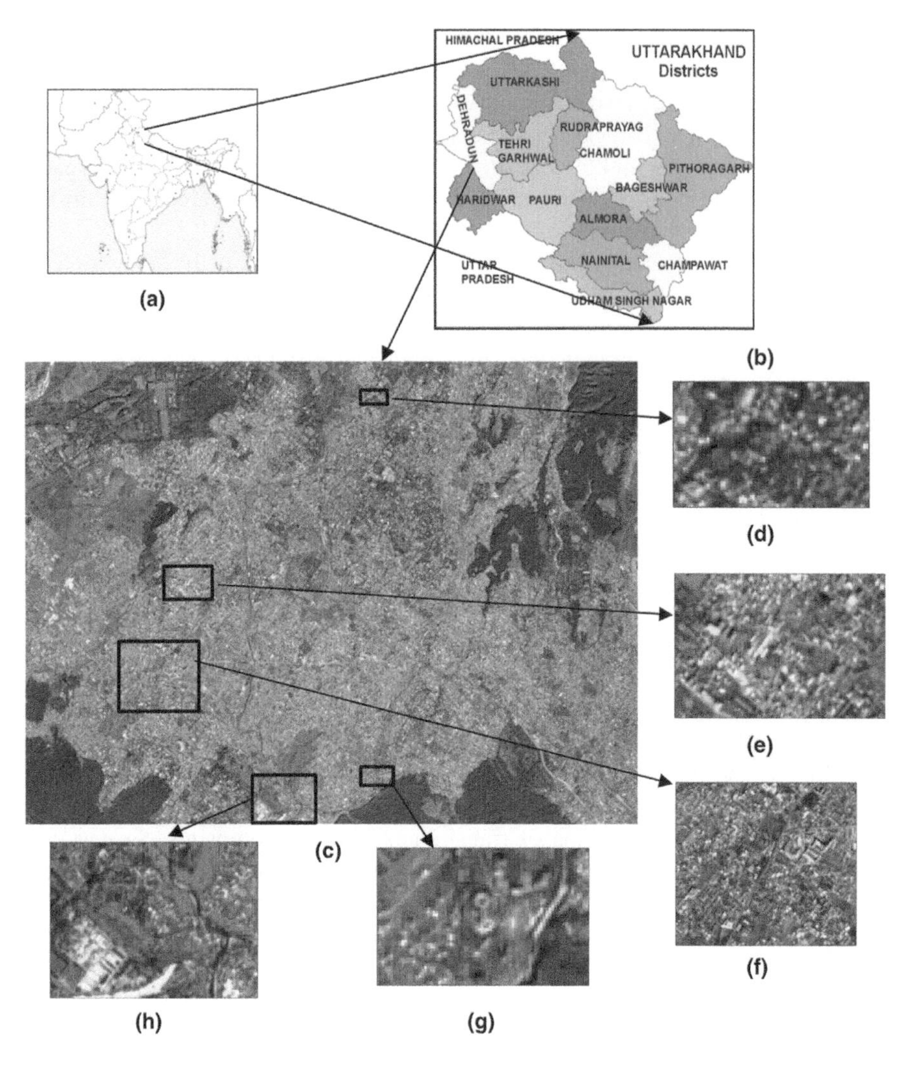

FIGURE A1.38a India map; **b** Uttarakhand districts; **c** Sentinel image: Dehradun; **d** Kalidas Road; **e** Sabji Mandi; **f** ISBT, Dehradun; **g** Doon University; **h** Training site: Bharuwala colony.

TABLE A1.19
Data specifications.

Satellite name	Sensor name	Spatial resolution	Spectral resolution
Landsat 8	Operational land imager (OLI), PAN	30 m, 15 m	7 optical bands
Sentinel 2A	Multi spectral instrument (MSI)	10 m	9 bands

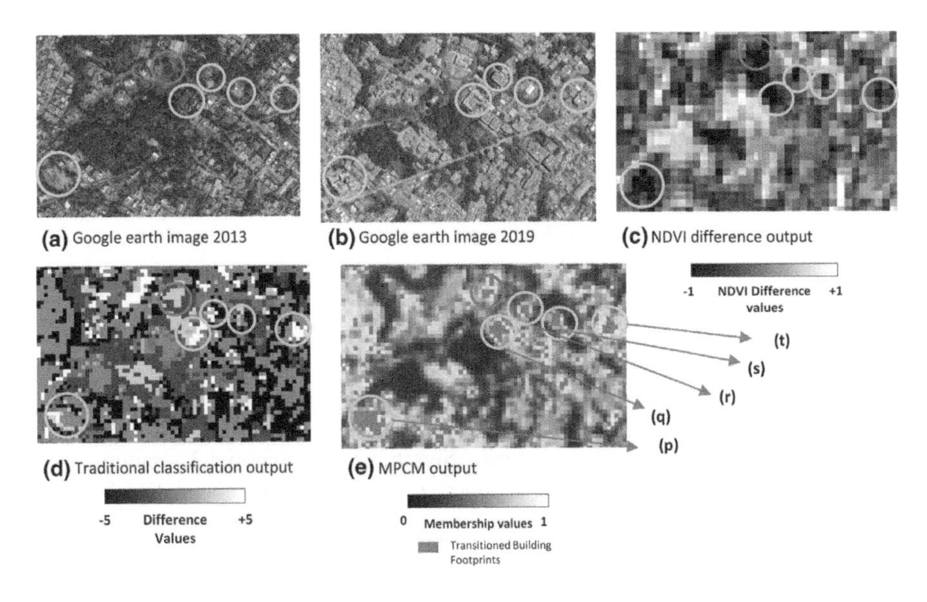

FIGURE 1.39 Kalidas Road: Google images: **a** 2013; **b** 2019; **c** NDVI difference; **d** post-classification change-detection; **e** fuzzy output.

transitioned-building footprints-extraction. High-resolution images can give better results in extracting small transitioned-building footprints such as residential houses and shops. Some noisy output pixels can be removed using Markov Random Field (MRF) or local-convolution methods, which add contextual information in the base classifier. Kalidas Road outputs generated with different methods are shown in Figure A 1.39.

In conventional change-detection techniques, multi-temporal satellite images are initially classified into respective land-cover classes; thereafter, these classified images were compared to detect the changes that occurred. This is a two-step process-i.e. classification followed with change detection. Using these conventional techniques, built-up change can be mapped from medium spatial-resolution remote-sensing data; however, it is not possible to map transitioned-building footprints. Secondly, in the case of coarse-resolution images, mixed pixels introduce classification errors, which are further propagated in the change-detection process. In the NDVI change-detection approach also, it was possible to map a vegetated area. But it is not feasible to map transitioned-building footprints from open areas. The present study proposes a fuzzy-based approach for detection of transitioned-building footprints from open spaces, while handling mixed pixels from coarse-resolution data. The study concludes that the mapping of the newly transitioned buildings' footprints from open spaces had been possible using a fuzzy machine-learning algorithm and temporal-indices data. In the case of the proposed fuzzy MPCM classification, the subjectivity is reduced, as it requires very few training samples and the algorithm selects the maximum and minimum reflectance-bands for the target class during indices generation, to reduce spectral dimensionality of multi-temporal images. Hence, the present study demonstrates the applicability of the proposed fuzzy machine-learning algorithm in urban areas where open spaces are rapidly transitioning to built-up areas. As future scope, detection of the transitioned-building footprints of smaller buildings can be carried out using a finer-spatial-resolution temporal data-set.

SUMMARY

This appendix includes all of the experiments conducted on specialised crop mapping, crop-harvesting monitoring, burnt paddy-field mapping, sunflower-sowing crop phases, medicinal crops such as Isabgol, and castor-crop mapping, as well as forest-species mapping. The use of dual-sensor Sentinel-1 and Sentinel-2 temporal data for crop and forest species mapping with a fuzzy model has also been discussed, as mapping paddy-transplanted crop-fields from multi-sensor, temporal, multi-spectral data with a 1D-CNN model.

BIBLIOGRAPHY

Attri, P., & Kushwaha, S. P. S. (2018). Estimation of biomass and carbon pool in barkot forest range, UK using geospatial tools. *ISPRS Annals of the Photogrammetry, Remote Sensing and Spatial Information Sciences*, 4(5), 121–128. https://doi.org/10.5194/isprs-annals-IV-5-121-2018

Chhapariya, K., Kumar, A., & Upadhyay, P. (2020). Handling non-linearity between classes using spectral and spatial information with kernel based modified possibilistic c-means classifier. *Geocarto International*. 37:6, 1704-1721, https://doi.org/10.1080/10106049.2020.1797186

Chhapariya, K., Kumar, A., & Upadhyay, P. (2021a). Kernel-based MPCM algorithm with spatial constraints and local contextual information for mapping paddy burnt fields. *Journal of the Indian Society of Remote Sensing*, 49. https://doi.org/10.1007/s12524-021-01346-1

Chhapariya, K., Kumar, A., & Upadhyay, P. (2021b). A fuzzy machine learning approach for identification of paddy stubble burnt fields. *Spatial Information Research*, 29, 319–329. https://doi.org/10.1007/s41324-020-00356-4

Congalton, R. G. (1991). A review of assessing the accuracy of classifications of remotely sensed data. *Remote Sens Environ*, 37(1), 35–46. https://doi.org/10.1016/0034-4257(91)90048 -B

Delegido, J., Verrelst, J., Luis, A., & Moreno, J. (2011). Evaluation of Sentinel-2 red-edge bands for empirical estimation of green LAI and chlorophyll content. *Sensors* (Basel), 11(7), 7063–7081. https://doi.org/10.3390/s110707063

Jensen, J. R. (1986). *Introductory Digital Image Processing: Aremote Sensing Perspective*. Prentice Hall PTR. United States.

Misra, G., Kumar, A., Patel, N. R., & Zurita-Milla, R. (2014). Mapping aspecific crop—a temporal approach for sugarcane Ratoon. *Journal of the Indian Society of Remote Sensing*, 42, 325–334. https://doi.org/10.1007/s1252 4-012-0252-1

Nünez, J., Otazu, X., Fors, O., Prades, A., Palà, V., & Arbiol, R. (1999). Multiresolution-based image fusion with additive wavelet decomposition. *IEEE Transactions on Geoscience and Remote Sensing*, 37(3I), 1204–1211. https://doi.org/10.1109/36.763274

Sokolova, M., Japkowicz, N., & Szpakowicz, S. (2006). Beyond accuracy F-score and ROC: A family of discriminant measures for performance evaluation. *AAAI Workshop Technical Reports*. Springer, Berlin, Heidelberg. https://doi.org/10.1007/11941439_114

Upadhyay, P., Ghosh, S. K., Kumar, A., Roy, P. S., & Gilbert, I. (2012). Effect on specific crop mapping using WorldView-2 multispectral add-on bands: Soft classification approach. *Journal of Applied Remote Sensing*, 6(1), 063524. https://doi.org/10.1117/1.JRS.6.063524

Experimentation capabilities for specific single class mapping from multi-sensor multi-temporal remote sensing images that are not available in commercial image processing software have been mentioned.

See Inside Shows Blessed With. . . .

See Outside Shows Missing Out. . . .

Appendix A2

SMIC—Temporal Data-Processing Module for Specific-Class Mapping

This appendix (A2) goes into detail about the capabilities of SMIC: sub-pixel multi-spectral image-classifier package for multi-temporal single-, dual-, or multi-sensor data-processing. The commercially available software has limitations for machine-learning algorithms capable of extracting specific single-class from multi-temporal remote-sensing images, even though a large number of classifiers, such as statistical-based linear-mixture model (LMM), fuzzy-set-based fuzzy c-means (FCM), and artificial neural network (ANN) have been implemented in commercial software. Limited algorithms have been incorporated in different commercially available digital-image-processing softwares like neural network and unsupervised, fuzzy *c*-Mean in PCI Geomatica and linear-mixture model (LMM) in ERDAS, ENVI, etc.

The SMIC package, on the other hand, is a menu-driven type with a large number of fuzzy-based, supervised soft-classifiers implemented under machine learning. Outputs from the SMIC temporal-data-processing module package can be generated as soft or hard classified-outputs and all parameters are open to do the experiment with different values assigned to all variables in fuzzy-based algorithms. Fuzzy- and deep-learning-based algorithms discussed in Chapters 5 and 6 have been implemented through the temporal module of an in-house developed package called SMIC (Figure A2.1).

The SMIC software has been developed using JAVA programming language and is specifically capable of soft/hard land-cover mapping from remote-sensing multi-spectral data. It has the capability of processing the multi-spectral mono/temporal remote-sensing satellite data at a sub-pixel level with various fuzzy-based classifiers. The system can handle multi-spectral images of any number of bands. In this package, various classification algorithms such as fuzzy *c*-means, possibilistic *c*-means, noise clustering, modified possibilistic *c*-means, and improved possibilistic *c*-means with or without entropy have been incorporated in supervised mode with soft/hard classification options. The second module of SMIC is developed to process multi-temporal data sets for specific single-class extraction (Figure A2.1).

In this module, the spectral dimensionality of temporal data is reduced through the class-based sensor dependent (CBSI) approach (Chapter 3, Section 3.2, and the temporal dimensionality can be maintained. Figure A2.2a provides an option to invoke the CBSI indices palate whereas Figure A2.2b provides an option for selecting mathematical indices expression while processing the temporal data. The advantage of CBSI is that the user does not have to specify band information from image data sets.

Reference-training data can be generated from this system in two modes: one is manually and the other is a region-growing concept. In manual mode, pixel-by-pixel training samples can be generated (Figure A2.3). In manual mode, the chances of

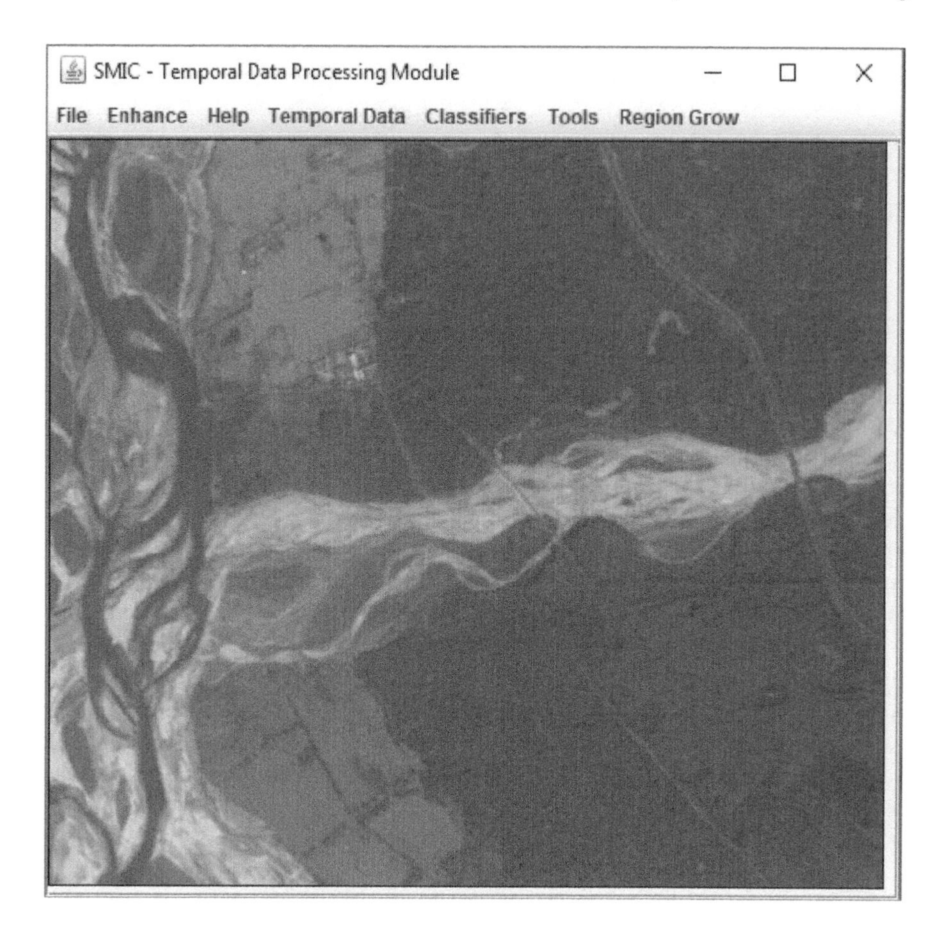

FIGURE A2.1 Temporal-data-processing module.

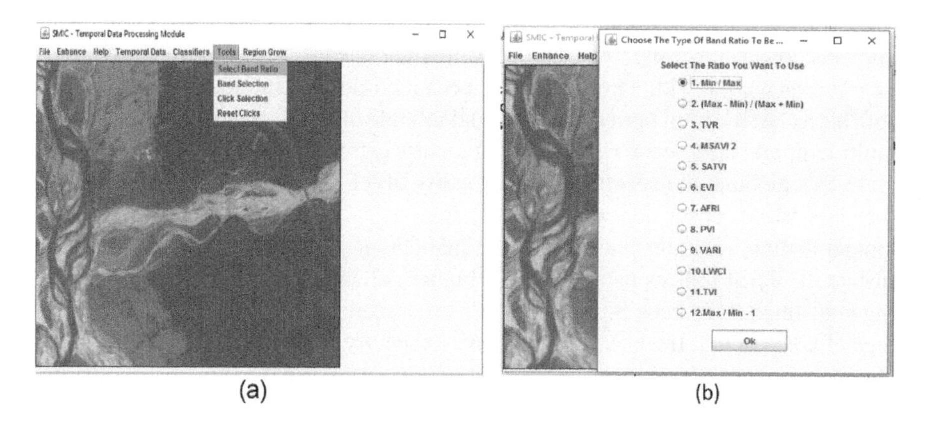

FIGURE A2.2 (a) and (b) show dimensionality-reduction methods.

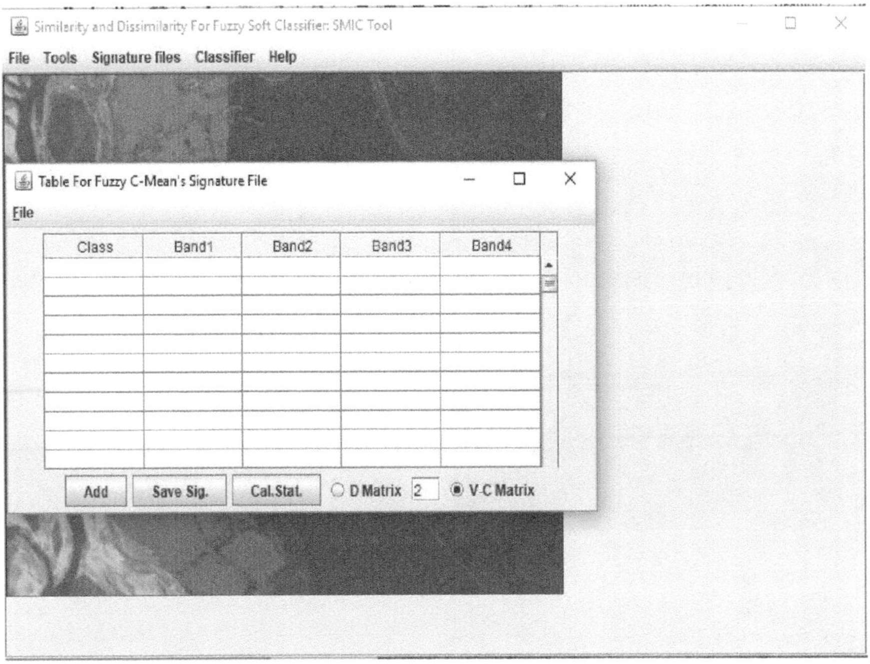

FIGURE A2.3 Manually training sample-collection module.

picking heterogeneous training samples are higher; hence, in such cases, region growing through seed pixels can be used for collecting training samples (Figure A2.4). Additionally, in region growing, similarity/dissimilarity algorithms have been implemented for generating the criteria to pick similar pixels in a given search window. Training-sample data of each class is saved in a separate training file. This region-growing method collects similar pixels as training-data sets using seed pixels while having various similarity/dissimilarity options. It also generates class-based sensor-independent indices using seed-pixel locations from temporal images.

Minimum human intervention has been considered while developing the multi-temporal data module-processing. This module includes temporal information as temporal-indices data while reducing the spectral information of multi-spectral images. The advantage of including temporal data is that it provides unique information about specific-class mapping due to reduced spectral overlapping with other classes. Figure A2.5 shows how to load other temporal images. Before loading other temporal images, first an image from the temporal database has to be loaded from a file option in Figure A2.1.

Figure A2.6 provides an option to save indices outputs generated. These indices layers generated from temporal data-sets give information as a temporal-indices database.

After generating the temporal-indices outputs, the next step is to apply a fuzzy-based classifier capable of identifying a specific class of interest. In this module,

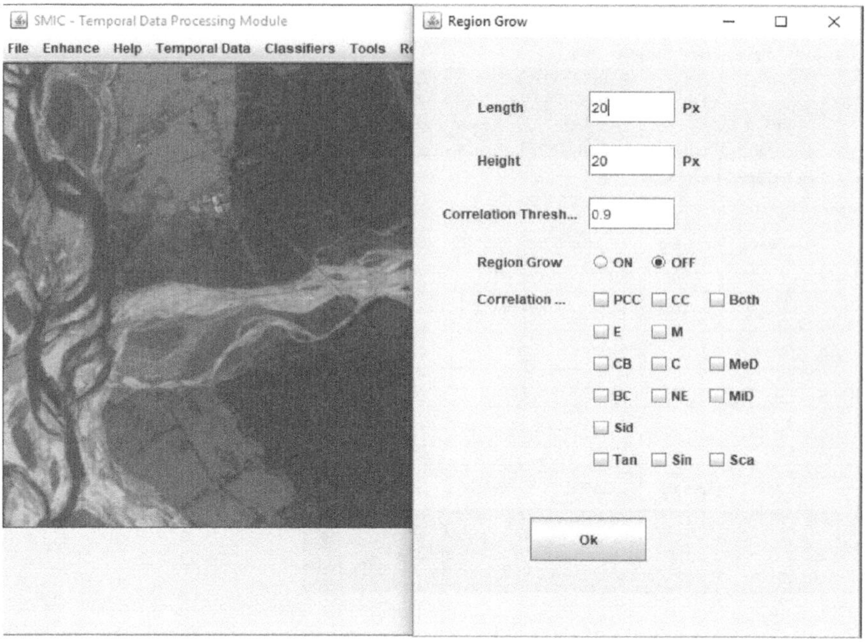

FIGURE A2.4 Region-growing methods.

PCM, NC, and MPCM classifiers (Chapter 5) have been implemented with various similarity/dissimilarity measures (Figure A2.7). The advantage of applying these classifiers is due to algorithms' capability of identifying the single class of interest with minimum parameters.

In a number of situations, when class size is small, it's quite difficult to collect a large number of training samples to train or calculate classification-algorithm parameters. The solution for this has been developed to increase training samples from small seed-training samples, while using statistical-constrained parameters generated from seed-training samples. The 'increasing training sample size' approach from seed pixels has been developed and included in the SMIC package. The 'increasing training sample size' approach through GUI is shown in Figure A2.8.

Another advantage of using the SMIC module for temporal-data processing is that the user needs just four steps to identify a specific class of interest using temporal data. Obtaining multi-temporal multi-spectral data from the same remote-sensing sensor is not always possible. In that case, this package has the capability of using the remote-sensing multi-spectral data available from different sensors. In addition, this module does not require knowledge about different spectral bands present in remote-sensing multi-spectral images. This module applies statistical operators to identify suitable bands to be used in indices generation and to have a temporal-indices database. Therefore, the overall advantage of this module is that technological knowledge is on the back end, and it requires minimum knowledge to work with temporal datasets for specific single-class mapping.

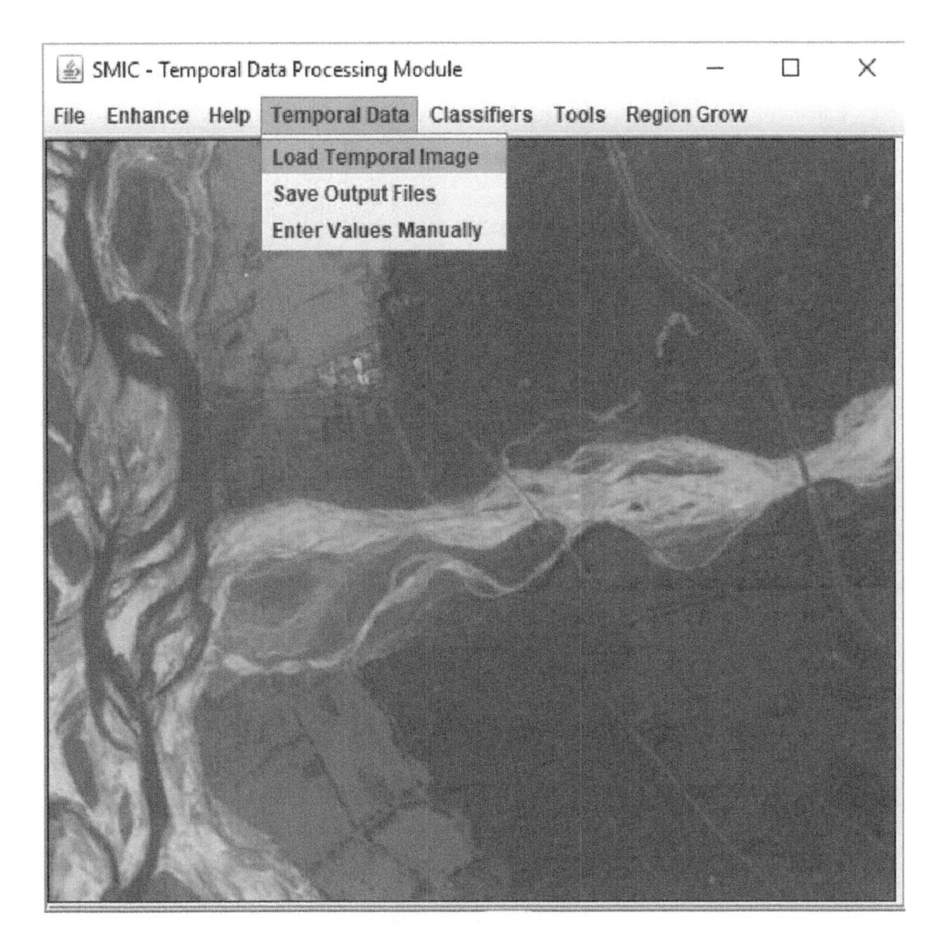

FIGURE A2.5 GUI shows how to load other temporal images.

While mapping a specific crop, for example, the knowledge of other crops present in that area is also important. Further, it is a fact that the spectral responses of different crops are likely similar to each other on a particular date and this makes the process complex for mapping specific crops using single-date imagery. Temporal information of a crop can, however, provide a unique information for discriminating among various crops and vegetation classes using the differences in their growth patterns as a discriminating factor. The need of multi-temporal data for continuous monitoring of crops and the unavailability of continuous temporal-data is also a well-known problem. Hence, a multi-sensor approach for increasing the multi-temporal data sampling for monitoring crops has to be evaluated for its effectiveness. SMIC is capable of handling the mono- as well as multi-sensor temporal images, and hence, is an effective package for specific-class identification and for mixed-pixel handling.

For deep learning models covered in Chapter 6, Pseudo code has been implemented as GUI, as shown in Figure A2.9. This GUI can be used for displaying

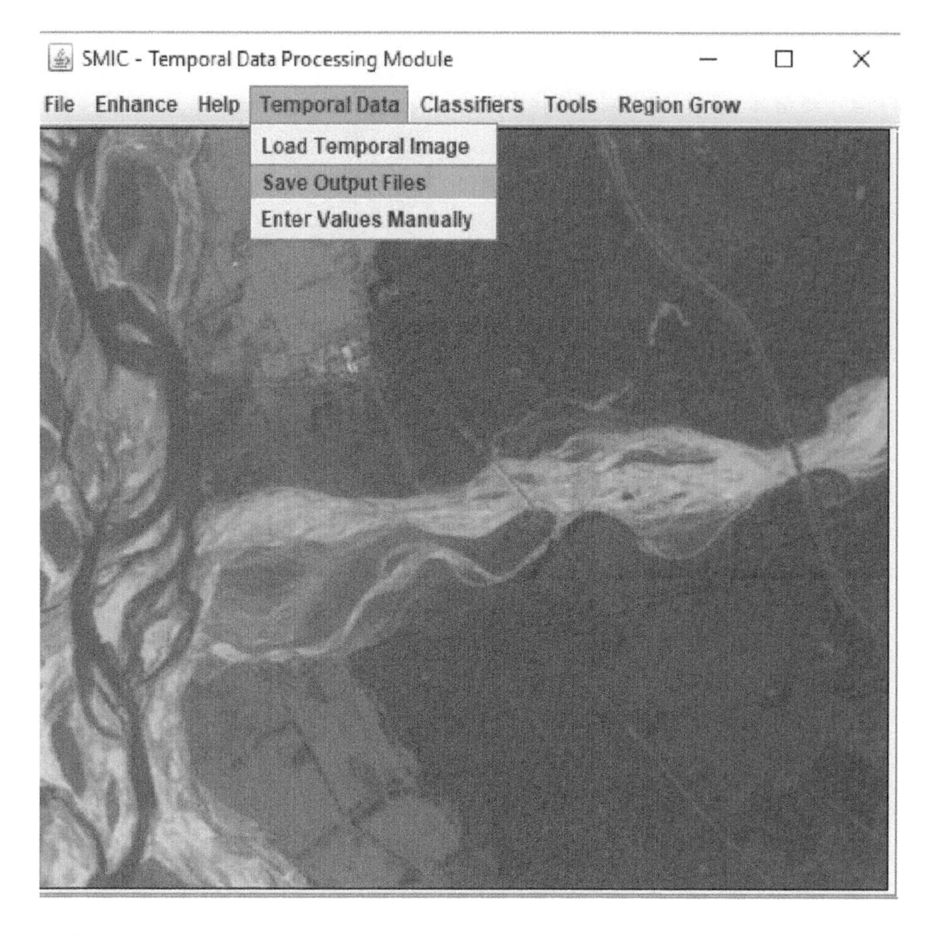

FIGURE A2.6 Saving indices outputs.

multi-spectral remote-sensing data. This GUI has the facility to collect reference data of various classes to be used for training and testing purposes (Figure A2.10). Reference data of each class is saved in different data files. In the next step, this GUI facilitates the reading and loading of training data of each class (Figure A2.11).

Once training data of each class has been loaded, the next step is to apply ANN or CNN networks through the training and testing module. By default, it gives an option of applying ANN with input images, called training data, defining ANN parameters like learning rate, epoch, and hidden neurons (Figure A2.12). While applying the CNN module, parameters of CNN have to be defined, such as epoch or threshold (Figure A2.13).

The CNN module provides options to add CNN-based layers, as well as to make it a hybrid approach while adding an LSTM layer (Figure A2.14). Once training is over, it will ask to save tuned weights of the model, as well as hard and soft classified

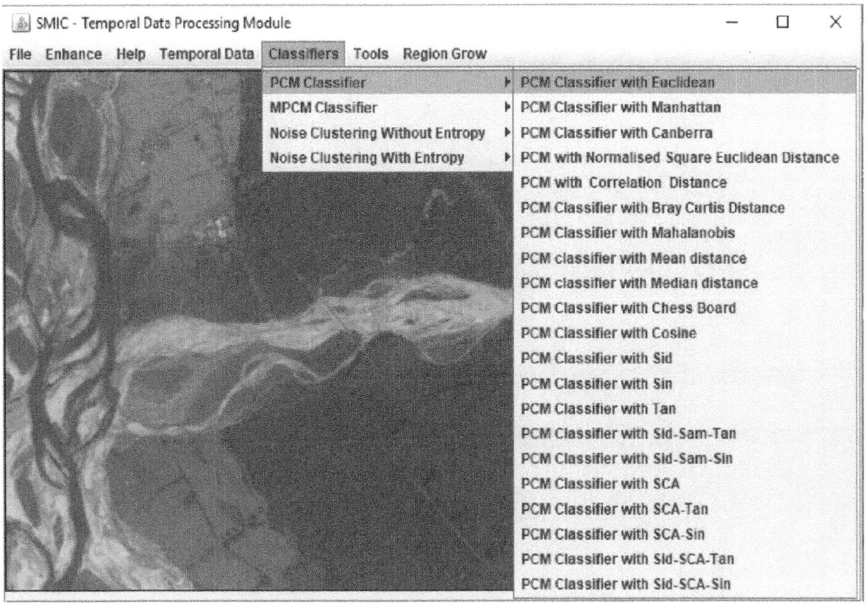

FIGURE A2.7 Classification methods for specific-class mapping.

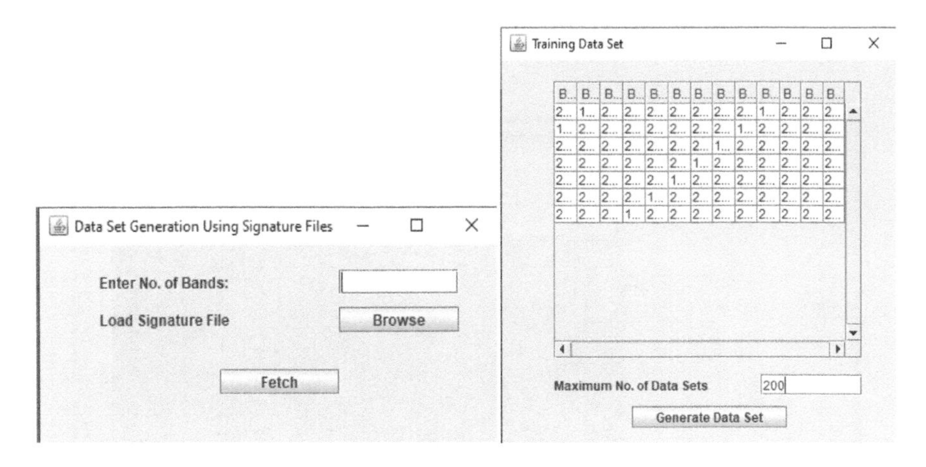

FIGURE A2.8 GUI to extend training-data size.

outputs. In hard classified output, it has been assumed that all pixels are pure pixels. In soft classified outputs, fraction images of each class are generated. During training and testing/classification, process accuracy and loss-function status can also be monitored.

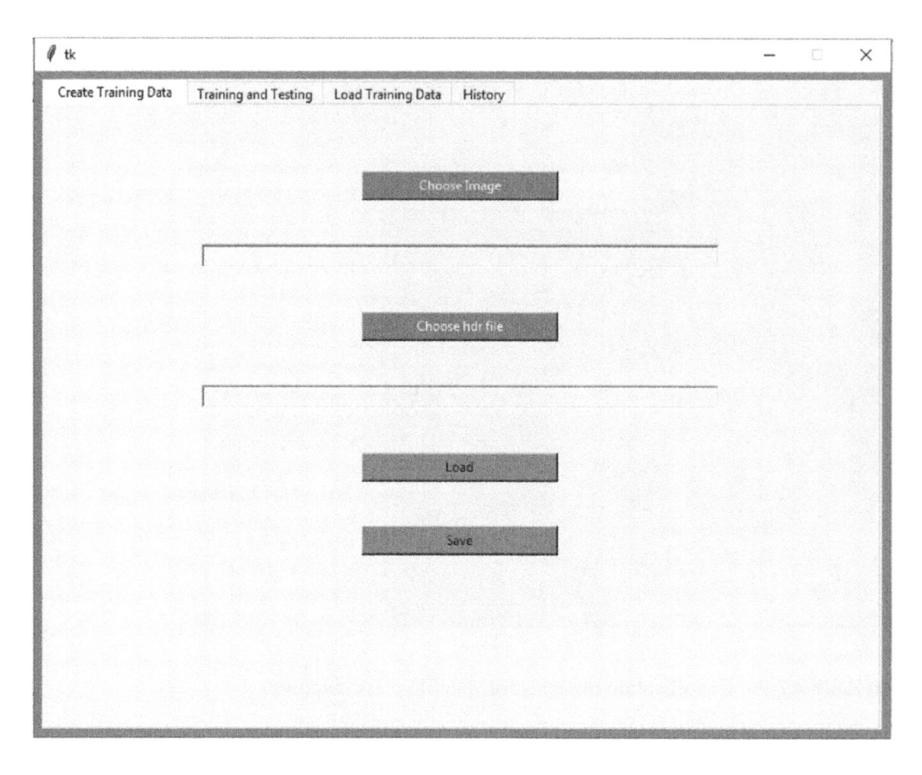

FIGURE A2.9 The main GUI of the learning tool.

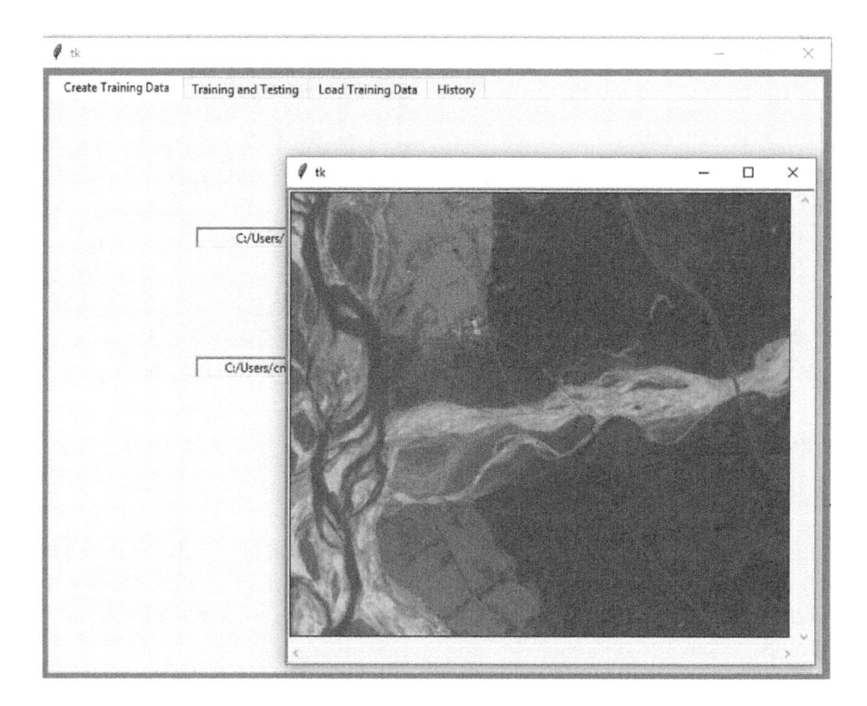

FIGURE A2.10 Training-data creation tool.

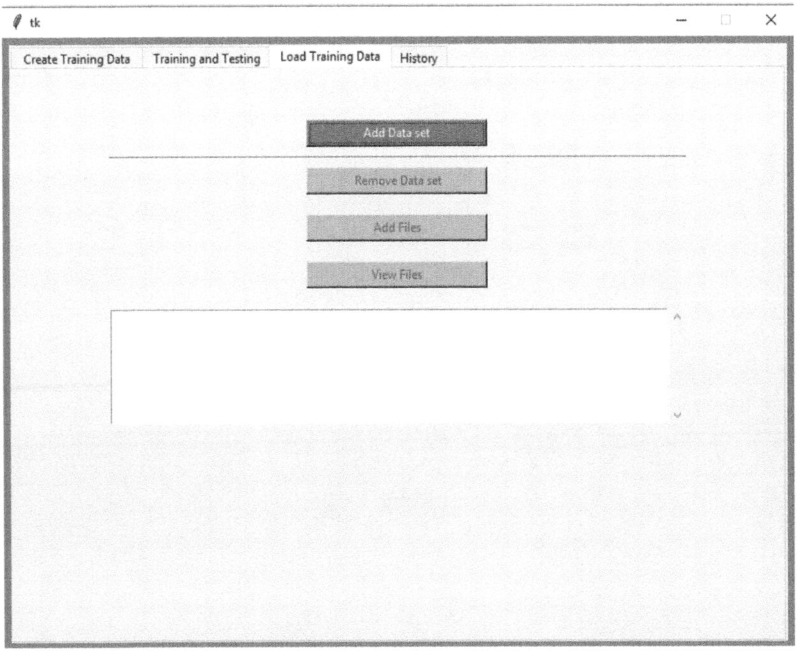

FIGURE A2.11 Load training-data module.

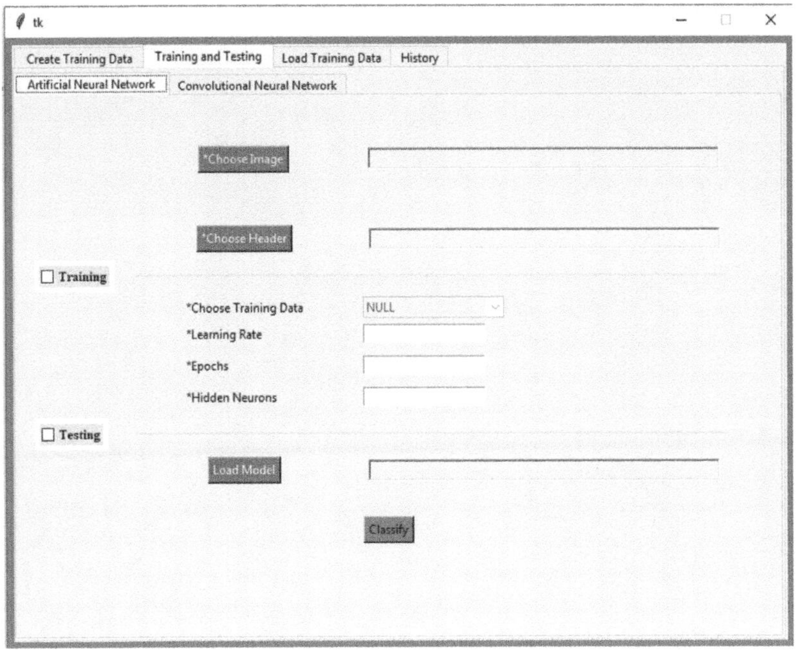

FIGURE A2.12 Training and testing module with ANN.

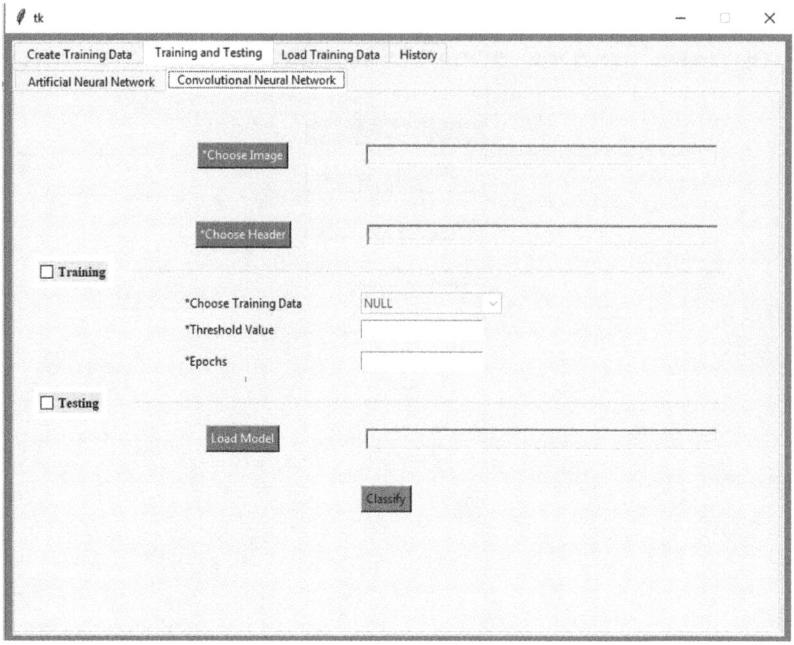

FIGURE A2.13 Training and testing module with CNN.

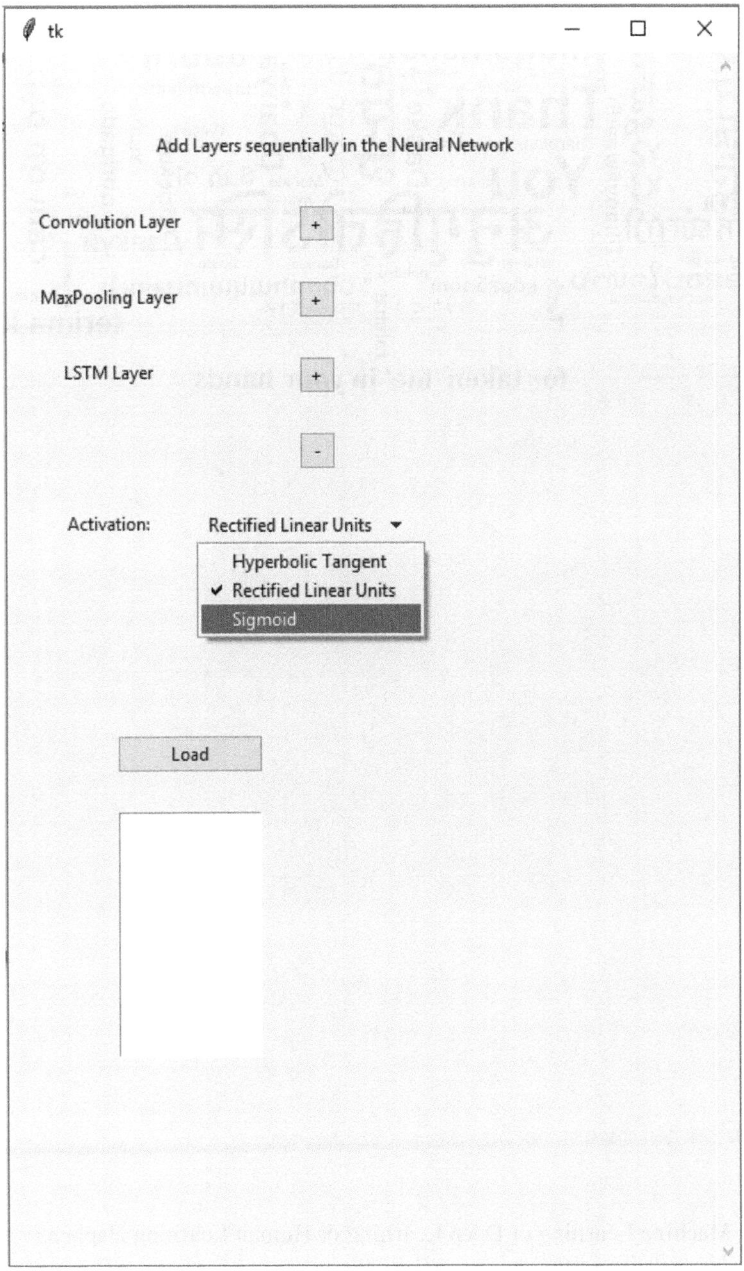

FIGURE A2.14 Module to add various layers in the CNN module.

Machine Learning or Deep Learning or Human Learning Happens Through Training. . . .

. . . .Comes Only From. . . .

Practice. . . .

Practice. . . .

Practice. . . .

Index

Note: Page numbers in **bold** and *italics* refer to tables and figures, respectively.